Cyanobacterial Nitrogen Metabolism and Environmental Biotechnology

Cyanobacterial Nitrogen Metabolism and Environmental Biotechnology

Edited by
Ashwani K. Rai

Springer-Verlag

Narosa Publishing House

Edited by
Ashwani K. Rai
Department of Botany
Banaras Hindu University
Varanasi 221 005, India

Exclusive distribution in North America, Europe, Latin America, Canada and Japan by Springer-Verlag Berlin Heidelberg New York.

For all other countries exclusive distribution by Narosa Publishing House, New Delhi.

This book has been produced from the Camera Ready Text provided by the Editor.

ISBN 3-540-61305-6 Springer-Verlag Berlin Heidelberg New York
ISBN 0-387-61305-6 Springer-Verlag New York Berlin Heidelberg
ISBN 81-7319-114-X Narosa Publishing House, New Delhi

Printed in India.

CONTRIBUTORS

HELMAR ALMON, *Botanisches Institut der Universität Bonn, Kirschallee 1, D-5300 Bonn 1, Germany*

IOAN I ARDELEAN, *Institute of Biology, Splaiul Independentei 296, Bucharest 79651, POB 17-50, Romania*

HERBERT BÖHME, *Botanisches Institut der Universität Bonn, Kirschallee 1, D-5300 Bonn 1, Germany*

SAMMY BOUSSIBA, *Microalgal Biotechnology, The Jacob Blaustein Institute for Desert Research, Ben-Gurion University of the Negev, Sede Boker campus, Israel 84990*

NORBERT ERDMANN, *University of Rostock, Department of Biology, Doberaner str. 143, D-18051, Germany*

ENRIQUE FLORES, *Instituto de Bioquímica Vegetal y Fotosíntesis, Universidad de Sevilla-CSIC, Facultad de Biología, Apartado 1113, E-41080 Sevilla, Spain*

MIGUEL G GUERRERO, *Instituto de Bioquímica Vegetal Y Fotosíntesis, Universidad de Sevilla-CSIC, Apdo. 1113, 41080- Sevilla, Spain*

MARTIN HAGEMANN, *University of Rostock, Department of Biology, Doberaner str. 143, D-18051, Germany*

ANTONIA HERRERO, *Instituto de Bioquímica Vegetal y Fotosíntesis, Universidad de Sevilla-CSIC, Facultad de Biología, Apartado 1113, E-41080 Sevilla, Spain*

CATALINA LARA, *Instituto de Bioquímica Vegetal Y Fotosíntesis, Universidad de Sevilla-CSIC, Apdo. 1113, 41080-Sevilla, Spain*

JACEK PLAZINSKI, *Research School of Biological Sciences, Australian National University, GPO Box 475, Canberra City, ACT 2601, Australia*

ASHWANI K RAI, *Department of Botany, Banaras Hindu University, Varanasi 221 005, India*

Preface

I must admit that the publication of NG Carr and BA Whitton *The biology of blue-green algae* and that of GE Fogg, WDP Stewart, P Fay and AE Walsby *The blue-green algae* in 1973 came as a great fortune and support to me at the time when I was to write my Ph.D. thesis. Until then there was not a single volume which could serve as a reference book or provide an overview of cyanobacterial biology. Later, several volumes became available: WDP Stewart's (partly devoted to cyanobacteria) *Algal physiology and biochemistry* (1974), *The biology of cyanobacteria* (1982) by NG Carr and BA Whitton, *Biochemistry of the algae and cyanobacteria* (1988) by LJ Rogers and JR Gallon, *The cyanobacteria* (1987) by P Fay and C van Baalen, *Cyanobacteria* (1988) of L Packer and AN Glazer and the latest volume edited by DA Bryant *The molecular biology of cyanobacteria* (1994). These volumes reflect tremendous progress in the area of cyanobacteriology.

In this context I would like to quote R.Y. Stanier: "the cyanobacteria, like a sleeping beauty, just emerging from a profound coma of 150 years. .... I can heartily recommend the leisurely exploration of hitherto neglected branch of prokaryotes. ....Almost any question one might ask, whether major or minor, could be answered by: I don't know" (Annu Rev Microbiol 1980, 34:1-48).

Cyanobacteria have now moved from the hands of phycologists to the agronomists, microbiologists, physiologists, biochemists and molecular biologists either as a convenient tool or as a unique organism harbouring higher plant type photosynthesis in a simple prokaryotic organization coupled with nitrogen fixing ability.

The concept for *Cyanobacterial Nitrogen Metabolism and Environmental Biotechnology* developed during discussions with my students while selecting their research topics. During discussions, we felt a lack of an up-to-date picture of the current state of various aspects of cyanobacterial nitrogen metabolism, because the literature is scattered in various journals and review articles. Also, the last decade has seen considerable advancement in research on cyanobacterial nitrogen metabolism and its potential exploitation in environmental sciences.

In this book, an attempt is made to compile the available information to provide genetical, ecophysiological, biochemical and biotechnological information in a single volume for graduate students and those working on cyanobacterial nitrogen metabolism. The book consists of chapters on cyanobacterial nitrate metabolism, ammonia assimilation, urea metabolism, nitrogen metabolism of symbiotic systems, interaction of carbon and nitrogen metabolism and environmental stresses, and their biotechnological and environmental aspects. In addition there is a chapter on methodologies used in the study of nitrogen metabolism and a chapter on future prospects. The book does not have a separate chapter on nitrogen fixation. Cyanobacterial nitrogen fixation is one of the most studied topics, hence, many decent review articles have been written on this topic including an up-to-date state-of the art. Therefore, this volume contains only the applied aspects of cyanobacterial nitrogen fixation.

This project would not have been possible without the help of many. My thanks must, first of all go to the contributors, each a recognized authority in his/her particular field, for their excellent contributions and patience. I also acknowledge the help of my students JP Singh and SP Tiwari who worked on my behalf on several occasions. I am grateful to my senior colleague, Professor JS Singh for his affection and moral support in time of need. Finally, I would like to thank my wife Krishna, my daughter Snigdha and my sons Abhisek and Amiya for allowing me the time of their share.

ASHWANI K RAI

Contents

Nitrate Metabolism

ANTONIA HERRERO and ENRIQUE FLORES

Instituto de Bioquímica Vegetal y Fotosíntesis, Universidad de Sevilla-CSIC, Facultad de Biología, Apartado 1113, E-41080 Sevilla, Spain

1 Introduction

Nitrate is the most abundant combined source of nitrogen available in Nature for plants and microorganisms, and nitrate assimilation is the most common mode of nitrogen nutrition in the cyanobacteria.

With very few exceptions, for example those of some thermophilic cyanobacteria (GARCÍA-PICHEL and CASTENHOLZ 1990), the cyanobacteria are able to grow with nitrate as the nitrogen source; thus, culture media commonly used in the laboratory include nitrate as a source of combined nitrogen (see e.g., RIPPKA et al. 1981). The concentration of nitrate found in natural environments is relatively low, for instance <0.01-50 µM in seawater or in some freshwater lakes (FOGG 1975, SYRETT 1981). As we shall see, nitrate is actively transported into the cyanobacterial cell, intracellular nitrate being photosynthetically reduced via nitrite to ammonium which is the inorganic form of nitrogen that is incorporated into organic compounds. We shall first present the characteristics of the cyanobacterial nitrate assimilatory system as deduced from physiological and biochemical studies. Then, the results of the recently developed genetic approach to this topic will be presented. Ammonium ions can also be found in natural environments (SYRETT 1981) and, in the laboratory, ammonium has been shown to behave as an effective inhibitor of nitrate assimilation in cyanobacteria. The physiological characteristics of this inhibition, which takes place both at the level of activity and at the level of genetic expression of the nitrate assimilatory system, as well as the regulatory effect of nitrate on the expression of this system in some cyanobacteria, will be presented. Finally, the genetic basis for the phenomenon of repression by ammonium will also be addressed.

2 Nitrate and nitrite uptake: the physiological evidence

The first step in the assimilation of any nutrient by a living cell is the entrance of that nutrient into the cell. For most nutrients, specialized transport systems operate their transport through the cell envelope. The structure of the cyanobacterial cell envelope is that found in the Gram-negative eubacteria, comprising a cytoplasmic membrane, a layer of peptidoglycan that can be rather thick, and an outer membrane (DREWS and WECKESSER 1982). The cytoplasmic membrane represents the main permeability barrier of the Gram-negative bacterial cell, since the outer membrane contains special proteins, known as porins, which form channels of sufficient diameter to pass molecules of up to ca. 600 molecular weight, including inorganic ions (OSBORN and WU 1980). There is evidence for the presence of such porins in the cyanobacterial cell wall (BENZ and BÖHME 1985). Therefore, nitrate probably has free access to the cyanobacterial periplasm.

2.1 Nitrate transport

In *Synechococcus* sp. PCC 6301 (formerly known as *Anacystis nidulans*), the transport of nitrate into the cell was suggested to be mediated by a transport system which could be defined by its relatively high affinity, as compared to that exhibited by nitrate reductase (FLORES et al. 1983a). Using different experimental approaches, the nitrate concentration required for half saturation of nitrate transport (K_S) in *Synechococcus* spp. has been found to be about 1-4 µM (TISCHNER and SCHMIDT 1984, OMATA et al. 1989, RODRÍGUEZ et al. 1992). A high-affinity transport system (K_S, 2 µM) has also been described in *Oscillatoria agardhii* (ZEVENBOOM and MUR 1979), a filamentous cyanobacterium that also exhibits a second uptake system with lower affinity for nitrate (K_S, 23-60 µM). In two strains of the genus *Anabaena*, a single component of nitrate uptake with a K_S for nitrate of 26.2 and 29.7 µM, respectively has been reported (MEEKS et al. 1983). N-starved cells of *Phormidium laminosum* exhibit a K_S for nitrate of 4.5 µM (GARBISU et al. 1992).

The unicellular strains *Synechococcus* sp. PCC 6301 and PCC 7942 have been shown to be able to concentrate nitrate inside the cell (LARA et al. 1987, RODRÍGUEZ et al. 1992). The intracellular level of nitrate which is reached during nitrate assimilation is determined by the rates of (i) inwards transport, (ii) reduction, and (iii) leakage from the cell (SHEARER et al. 1991). Considering together the anionic character of nitrate, the fact that a negative membrane potential of about -100 mV is found in cyanobacteria (see e.g., PASCHINGER 1977), and the intracellular levels of nitrate that were determined experimentally, positive free-energy changes have been calculated for nitrate transport (LARA et al. 1987, RODRÍGUEZ et al. 1992). The endergonic character of this process is also manifest in the observed sensitivity of nitrate uptake to the uncoupler FCCP (OHMORI et al. 1977, RAI et al. 1981, FLORES et al. 1983a) and to the inhibitor of the F_0F_1-ATPase, DCCD (RAI et al. 1981, FLORES et al. 1983a, BAGCHI et al. 1989). The question of the mechanism of coupling of metabolic energy to nitrate transport can then be raised. Inhibition by DCCD indicates that ATP should participate at some step between the generation of metabolic energy and its use to drive nitrate transport.

Bacterial active transport systems can be broadly divided into two classes: (i) those which are usually composed of a single, very hydrophobic membrane protein and are energized by electrochemical gradients, and (ii) those which are composed by several proteins which constitute a membrane-bound complex, together with a periplasmic protein which is usually soluble. These "multicomponent", "periplasmic", or "binding protein-dependent" transport systems have recently been demonstrated to directly use ATP as a powering substrate (BISHOP et al. 1989; evidence reviewed by AMES and JOSHI 1990). Based on the positive effect of sodium ions on the level of accumulation of nitrate in a nitrate reductase mutant of *Synechococcus* sp., it has been suggested that a cotransport of nitrate and sodium with a downhill sodium electrochemical gradient from outside to inside the cells might be responsible for the active transport of nitrate (RODRÍGUEZ et al. 1992). However, as it will be discussed below, the genetic structure of the *Synechococcus* nitrate transport system has been recently unravelled and found to be that of a multicomponent transport system (OMATA et al. 1992). Indeed, a strict requirement for sodium ions for nitrate uptake to take place has not been demonstrated, and the observed influence of sodium on the extent of nitrate accumulation (RODRÍGUEZ et al. 1992) can be rationalized in terms of sodium having an effect on membrane potential which in turn might affect a process distinct from nitrate inwards transport, for instance nitrate efflux.

The occurrence of a high-affinity, concentrative transport system for nitrate in the cyanobacteria will permit the uptake by these organisms of the rather low concentrations of nitrate that may be found in their natural environments, and the intracellular concentration of nitrate reached will allow nitrate reductase to work. However, hampering of nitrate transport by inactivation of some nitrate transport genes (see below) do not preclude nitrate-dependent growth when nitrate is supplied at a high concentration, >20 mM (OMATA et al. 1989, 1992). Entrance of nitrate into the cell under these conditions appears to take place by passive diffusion since, at nitrate concentrations higher than 1 mM, nitrate uptake increases linearly as the nitrate concentration is increased (OMATA et al. 1989).

A cytoplasmic-membrane protein of about 48 kDa has been identified as a component of the nitrate transport system in *Synechococcus* sp. PCC 7942 based on its mode of regulation which is similar to that of other components of the nitrate assimilatory system, as well as on its altered expression in some pleiotropic nitrate-assimilation mutants (MADUEÑO et al. 1988b). Indeed, a mutant specifically lacking this protein was impaired in nitrate transport (OMATA et al. 1989). A similar protein is present in the cytoplasmic membrane of *Synechococcus* sp. PCC 6301 (SIVAK et al. 1989). As it will be discussed below, this protein might function as a binding protein for a multicomponent nitrate transport system.

2.2 Nitrite uptake

Two different components contribute to the uptake of low (e.g., 100 μM or less) concentrations of nitrite in a number of cyanobacterial strains (FLORES et al. 1983a, 1987, MARTÍN-NIETO et al. 1989): (i) an active transport of nitrite which is sensitive to DCCD; and (ii) a passive influx of nitrous acid, which is insensitive to DCCD and whose contribution to net nitrite uptake decreases as the pH of the medium is raised. K_S values of 6-22 μM for the active transport of nitrite have been reported (FLORES et al. 1987, MADUEÑO et al. 1987, GARBISU et al. 1992). Competitive interactions between nitrate and nitrite transport suggest that the same transport system is involved in the active transport of both substrates (MADUEÑO et al. 1987, RODRÍGUEZ et al. 1992).

3 Nitrate reduction

Intracellular nitrate is reduced to the term of ammonium which is the inorganic form of nitrogen that is incorporated into carbon skeletons. Nitrate reduction to ammonium takes place in two successive steps: (i) the two-electron reduction of nitrate to nitrite catalyzed by nitrate reductase; and (ii) the six-electron reduction of nitrite to ammonium catalyzed by nitrite reductase. In the cyanobacteria, both reductases use reduced ferredoxin as the physiological electron donor (HATTORI and MYERS 1967, MANZANO et al. 1976, ORTEGA et al. 1976, CANDAU et al. 1980, MÉNDEZ et al. 1981, Ida and MIKAMI 1983, ARIZMENDI and SERRA 1990, BLANCO and SERRA 1992). This has been demonstrated in *in vitro* assays where ferredoxin was reduced (i) with NADPH and ferredoxin-NADP oxidoreductase, (ii) photosynthetically with cyanobacterial thylakoids or green chloroplasts, (iii) with illuminated deazaflavin, or (iv) chemically with dithionite. However, dithionite had to be used in low concentrations in order to avoid inactivation of nitrate reductase (MIKAMI and IDA 1986; see GUERRERO and LARA 1987 and references therein). An alternative physiological donor for cyanobacterial nitrate reductase (CANDAU 1979, MIKAMI and IDA 1984) and nitrite reductase (MANZANO 1977, PESCHEK 1979) is flavodoxin, a protein that is expressed during growth under iron-limited conditions.

3.1 Nitrate reductase

The cyanobacterial nitrate reductase is a protein of 58-85 kDa (CANDAU 1979, MIKAMI and IDA 1984, BLANCO and SERRA 1992, MARTÍN-NIETO et al. 1992) which appears to be constituted by a single polypeptide (CANDAU 1979, MIKAMI and IDA 1984, BLANCO and SERRA 1992).

Like in all other nitrate-assimilating organisms, nitrate reduction in the cyanobacteria requires molybdenum (PESCHEK 1979, HERRERO and GUERRERO 1986, MARTÍN-NIETO et al. 1990). As it is generally true for molybdenum-containing enzymes, molybdenum will be present in the cyanobacterial nitrate reductase in the form of a molybdenum cofactor (BAGCHI et al. 1987). In addition to molybdenum, the cyanobacterial nitrate reductase contains non-heme iron and acid-labile sulfide (CANDAU 1979, MIKAMI and IDA 1984). A stoichiometry of one molybdenum, four non-heme irons, and four acid-labile sulfurs per molecule of nitrate reductase has been determined for the enzyme of *Plectonema boryanum* which might bear two Fe_2S_2 clusters (MIKAMI and IDA 1984).

In the reaction catalyzed by cyanobacterial nitrate reductase, two mol of ferredoxin are oxidized per mol of nitrate reduced (CANDAU 1979, MIKAMI and IDA 1984). The K_m for reduced ferredoxin has been determined for the nitrate reductase isolated from several cyanobacteria, showing values of 14 to 65 μM depending on the strain and the method used to reduce ferredoxin (CANDAU 1979, MIKAMI and IDA 1984, BLANCO and SERRA 1992). *Synechococcus* nitrate reductase exhibits a K_m for flavodoxin of 44 μM (CANDAU 1979). Artificial electron donors such as reduced methyl viologen (MANZANO et al. 1976, MIKAMI and IDA 1984) and reduced bromophenol blue (MARTÍN-NIETO et al. 1992) can also be used to assay cyanobacterial nitrate reductase.

The nitrate reductase assay originally described by MANZANO et al. (1976) which makes use of dithionite-reduced methyl viologen as an electron donor, has demonstrated to work in a wide number of cyanobacterial strains (HERRERO et al. 1981, 1985, MARTÍN-NIETO et al. 1989, 1992). Cyanobacterial cells made permeable with toluene (FLORES et al. 1980) or with the ionic detergent mixed alkyltrimethylammonium bromide (HERRERO et al. 1985) can be used as a source of enzyme. Because of the peculiarity of this assay, whose optimum pH is 10.5 (MANZANO et al. 1976, MIKAMI and IDA 1984), it can be postulated that nitrate reductases from very different cyanobacteria are probably very similar enzymes.

The K_m for nitrate in the methyl viologen-dependent nitrate reductase assay is of 0.7 to 3.5 mM for nitrate reductase from organisms like *Synechococcus* spp., *Plectonema boryanum* or *Phormidium laminosum* (CANDAU 1979, IDA and MIKAMI 1983, BLANCO and SERRA 1992, MARTÍN-NIETO et al. 1992). A peculiarity found for nitrate reductase from all the heterocystous cyanobacteria investigated to date, also shared by the enzyme from the unicellular strain *Synechocystis* sp. PCC 6803, is that a biphasic kinetic behaviour with respect to nitrate as the variable substrate is evident, with K_m values of 0.05-0.25 and 5-25 mM for the higher- and lower-affinity kinetic components, respectively (MARTÍN-NIETO et al. 1992). A single enzyme species appears to be responsible for the observed kinetic behaviour.

K_m values for nitrate of 55 and 65 µM were found when nitrate reductase from the filamentous, non-heterocyst forming cyanobacteria *Plectonema boryanum* and *Phormidium laminosum*, respectively, were assayed with reduced ferredoxin as a source of electrons (IDA and MIKAMI 1983, BLANCO and SERRA 1992).

3.2 Nitrite reductase

The cyanobacterial nitrite reductase consists of a single polypeptide of 52-68 kDa (HATTORI and UESUGI 1968, MANZANO 1977, MÉNDEZ and VEGA 1981, YABUKI et al. 1985, ARIZMENDI and SERRA 1990). Spectrophotometric studies suggested the presence of a heme group, probably a siroheme (MÉNDEZ and VEGA 1981, ARIZMENDI and SERRA 1990), as a prosthetic group of the enzyme. Analysis of the amino acid sequence of the enzyme (see below), which was deduced from the nucleotide sequence of the *nir* gene encoding nitrite reductase in *Synechococcus* sp. PCC 7942 (LUQUE et al. 1993), supports that the active center of the cyanobacterial nitrite reductase consists of an Fe_4S_4 cluster and a siroheme bridged by a cysteine sulfur. In the case of *Synechococcus* sp. PCC 7942, the enzyme is composed of 512 amino acids with a deduced MW of 56,502.

K_m of cyanobacterial nitrite reductase for reduced ferredoxin is 3.5-22 µM (MANZANO 1977, MÉNDEZ et al. 1981, ARIZMENDI and SERRA 1990). Nitrite reductase from a number of cyanobacteria can also be conveniently assayed with dithionite-reduced methyl viologen as an artificial electron donor (GUERRERO et al. 1974, MÉNDEZ et al. 1981, MARTÍN-NIETO et al. 1989, ARIZMENDI and SERRA 1990). K_m for nitrite of nitrite reductase, using methyl viologen as an electron donor, ranges from 40 to 110 µM depending on the source of the enzyme (MANZANO 1977, MÉNDEZ et al. 1981, ARIZMENDI and SERRA 1990).

3.3 *In situ* and *in vivo* nitrate reduction

Both reductases are found firmly bound to thylakoid membranes when subcellular preparations of *Synechococcus* sp. are isolated (GUERRERO et al. 1974, MANZANO et al. 1976). In *Anabaena cylindrica* and *Nostoc muscorum (Anabaena* sp. PCC 7119), the same is true for nitrate reductase (HATTORI and MYERS 1967, ORTEGA et al. 1976) but not for nitrite reductase which is soluble or only loosely bound to membranes (HATTORI and MYERS 1966, MÉNDEZ et al. 1981). On the other hand, both reductases have been described as soluble proteins in *Phormidium laminosum* (ARIZMENDI and SERRA 1990, BLANCO and SERRA 1992).

Attachment of nitrate reductase and nitrite reductase to thylakoid membranes permits the photoreduction of nitrate (HATTORI and MYERS 1967, MANZANO et al. 1976, ORTEGA et al. 1976, PESCHEK 1979) or nitrite (GUERRERO et al. 1974, MANZANO et al. 1976, PESCHEK 1979) in a Photosystem I-dependent reaction which is stimulated by added ferredoxin (HATTORI and MYERS 1967, MANZANO et al. 1976, ORTEGA et al. 1976). Notably, with *Synechococcus* thylakoids, photoreduction of nitrate to ammonium at the expenses of water as a source of electrons was observed in a reaction involving both Photosystem I and Photosystem II (CANDAU et al. 1976). Photolysis of water coupled to the reduction of nitrate to nitrite by *Anabaena* thylakoids has also been reported (ORTEGA et al. 1976).

The use by cyanobacterial nitrate reductase and nitrite reductase of reduced ferredoxin as an electron donor, and the coupling of nitrate reduction to oxygenic photosynthesis in thylakoid membranes of some cyanobacteria, as well as the inhibition by DCMU of *in vivo* nitrate reduction reported for *Synechococcus* sp. and *Oscillatoria chalybea* (FLORES et al. 1983a, BEDNARZ and SCHMID 1991), suggest that nitrate reduction is, in these organisms, a genuine photosynthetic process, analogous to the photosynthetic reduction of NADP. The operation of nitrate reduction *in vivo* as a photosynthetic process was demonstrated by the quantitation of the oxygen evolution associated to nitrate uptake and reduction up to the term of ammonium (FLORES et al. 1983a). In an experimental system involving the use of *Synechococcus* sp. PCC 6301 cells treated with L-methionine- D,L-sulfoximine (MSX), which do not assimilate the ammonium resulting from nitrate (or nitrite) reduction because of the inactivation by MSX of glutamine synthetase (see below), stoichiometries of 2 and 1.5 mol of oxygen evolved per mol of nitrate and nitrite, respectively, reduced to ammonium are found (FLORES et al. 1983a). These are the figures expected if the 8 and 6 electrons required for the reduction of nitrate and nitrite, respectively, to ammonium originated from the photolysis of water.

In agreement with the photosynthetic nature of nitrate reduction, quenching of chlorophyll fluorescence by nitrate and nitrite has been documented in *Anabaena* sp. (SERRANO et al. 1981, 1982), and the action spectra for both nitrate and nitrite assimilation has been shown to be coincident with the absorption band(s) of c-phycocyanin in *Synechococcus* sp. or of c-phycoerythrin and c-phycocyanin in *Calothrix* sp. (SERRANO and LOSADA 1988). This demonstrates that phycobiliproteins, closely associated *in vivo* to Photosystem II, are actively involved in photoreduction of both nitrate and nitrite in the cyanobacteria.

The question of how some cyanobacteria are able to grow heterotrophically in the dark using nitrate as a nitrogen source can be raised. For nitrate reduction under these conditions, it could be postulated that ferredoxin reduced by NADPH and ferredoxin-NADP oxidoreductase would probably be used as a source of electrons, NADPH being generated by glucose-6-phosphate dehydrogenase and 6-phosphogluconate dehydrogenase. Generation of NADPH in these reactions has also been claimed to explain DCMU- insensitive nitrate reduction to nitrite in *Phormidium uncinatum* (BAGCHI et al. 1989).

4 Incorporation of ammonium into carbon skeletons

A brief discussion of the fate of the ammonium resulting from nitrate and nitrite reduction is appropriate. In cyanobacteria, in general, ammonium assimilation proceeds mostly via the cycle glutamine synthetase/glutamate synthase, with possibly some minor contributions of other enzymes like alanine dehydrogenase or glutamate dehydrogenase (WOLK et al. 1976, MEEKS et al. 1977, 1978).

The cycle renders glutamate which is the principal nitrogen donor for the synthesis of other organic nitrogenous compounds and is found in the cyanobacteria at very high levels (see e.g., CORONIL and LARA 1991). Operation of the glutamine synthetase/ glutamate synthase pathway in the cyanobacteria is effectively hampered by MSX, an irreversible inhibitor of glutamine synthetase (see e.g., STEWART and ROWELL 1975, WOLK et al. 1976, RAMOS et al. 1982, FLORES et al. 1983b), or by substances like azaserine which are inhibitors of the glutamine amido transferases (WOLK et al. 1976, FLORES et al. 1983b). Cyanobacterial cells treated with MSX and fed with nitrate or nitrite release to the outer medium the ammonium generated by nitrate or nitrite reduction (RAMOS et al. 1982, FLORES et al. 1983a, RAMOS and GUERRERO 1983); in *Synechococcus* sp. PCC 6301, an amount of ammonium close to 90% of the nitrate or nitrite taken up is found to be released (RAMOS et al. 1982, FLORES et al. 1983a). Glutamine synthetase is an ATP-requiring enzyme. When *Synechococcus* cells treated with the ATPase inhibitor DCCD are supplied with nitrite, 90% of the nitrogen of the nitrite taken up (nitrite uptake under these conditions takes place as passive diffusion, see above) is recovered as ammonium in the outer medium (FLORES and GUERRERO, unpublished results). In *Synechococcus* sp., inhibition of glutamate synthase with azaserine also results in release to the outer medium of the ammonium generated by nitrate reduction, in an amount accounting for 85% of the nitrate taken up (FLORES et al. 1980). These results point to a central role of the glutamine synthetase/glutamate synthase pathway in the assimilation of the ammonium generated intracellularly by nitrate or nitrite reduction.

A summary of some aspects of the first steps of nitrate assimilation in cyanobacteria is presented in Fig. 1.

5 Genetic structure of the system for nitrate assimilation

A region of the genome of the unicellular cyanobacterium *Synechococcus* sp. PCC 7942 has been cloned (LUQUE et al. 1992, OMATA et al. 1993) in which a cluster of genes encoding proteins involved in nitrate assimilation has been located. The genes that have

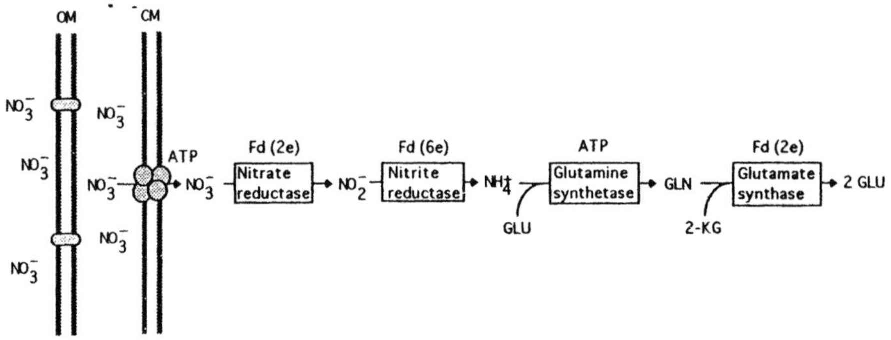

Fig. 1. First steps of nitrate assimilation in cyanobacteria. Nitrate enters the cell through a multicomponent transport system, being then reduced to the term of ammonium which is incorporated into carbon skeletons. Reduced ferredoxin (a one-electron carrier) is the physiological electron donor for nitrate reductase, nitrite reductase, and glutamate synthase. ATP is used by the nitrate permease and by glutamine synthetase. OM, outer membrane; CM, cytoplasmic membrane.

been identified in this cluster, as well as the characteristics of the proteins encoded by them, as deduced from their derived amino acid sequences, are described in the following paragraphs.

5.1 The *nir* gene

The *nir* gene has been identified as the structural gene for nitrite reductase by means of the study of the phenotype exhibited by mutant strains bearing transposon or gene-cassette insertions in that gene, that exhibited undetectable levels of nitrite reductase (LUQUE et al. 1992), and by sequence analysis (LUQUE et al. 1993). This showed that the *nir* gene would encode a protein (Nir) of 56,502 Da that exhibits 54 and 50% identity to nitrite reductase from *Zea mays* (maize) and *Spinacea oleracea* (spinach), respectively. Nir from *Synechococcus* sp. also shows homology to the hemoprotein component of sulfite reductase from *Escherichia coli* and *Salmonella typhimurium*. Identical amino acid residues are distributed along the entire sequences for the five mature proteins. Nir from *Synechococcus* sp. (LUQUE et al. 1993) bears the four conserved cysteine residues, as well as other conserved amino acids surrounding those cysteines (namely, the sequences Cys-$(X)_5$-Cys-$(X)_2$-Ala-$(X)_2$-Glu, and Thr-Gly-Cys-Pro-Asn-Gly/ Ser-Cys-Gly), that have been attributed structural importance in the configuration of the active site of the sulfite reductase hemoprotein (OSTROWSKI et al. 1989) and that are also present in higher-plant nitrite reductase. This strongly suggests that, as it is the case for enterobacterial sulfite reductase hemoprotein and higher-plant nitrite reductase (OSTROWSKI et al. 1989), nitrite reductase from *Synechococcus* sp. also bears a catalytic site constituted by an Fe_4S_4 cluster and a siroheme in which the Fe_4S_4 cluster is bound to the protein through the above-mentioned four conserved cysteine residues, with the sulfur of one of them bridging the Fe_4S_4 cluster to the siroheme.

5.2 The *nrt* genes

The gene *nrtA* (OMATA 1991) encoding the 48 kDa polypeptide involved in nitrate transport was cloned by screening with antibodies an expression gene bank of *Synechococcus* sp. PCC 7942 (OMATA et al. 1989), and has been located downstream of the *nir* gene (LUQUE et al. 1992). Downstream of *nrtA* three other genes (*nrtBCD*) have been identified by sequencing (OMATA et al. 1993) whose deduced products show homology to the subunits of bacterial transport systems dependent on binding proteins. Inactivation of any of these genes with antibiotic-resistance gene-cassettes results in an impairement in nitrate transport (OMATA et al. 1993). NrtA is a rather hydrophilic protein which, although being membrane-bound, has been proposed to represent the substrate-binding protein of the *nrt* system (OMATA et al. 1993).

NrtB is a hydrophobic protein of 279 amino acids which bears five (or may be six) hydrophobic amino acid segments that can span the membrane (OMATA et al. 1993). NrtB is homologous to ProW, the integral membrane component of the glycine betaine permease of *Escherichia coli*. Because the membrane protein complex of transport systems generally contains 10 or 12 membrane-spanning segments (MALONEY 1990), and because *nrtB* is the only gene of the nitrate assimilation cluster encoding a very hydrophobic protein, it might be anticipated that NrtB would act as a homodimer.

NrtD is a protein of 274 amino acids which bears ATP-binding motifs and exhibits extensive homology to the "conserved components" of the bacterial multicomponent transport systems (OMATA et al. 1993). The "conserved component" protein (AMES 1986) represents the site of ATP binding and hydrolysis and, although being hydrophilic, it forms a complex with the hydrophobic proteins of the system (KERPPOLA et al. 1991).

NrtC is a polypeptide of 659 amino acids that bears a large C-terminal domain with structural similarity to NrtA (30% identity) and an N-terminal domain that has features of an ATP-binding domain and is homologous (58% identity) to NrtD (OMATA et al. 1993). Several multicomponent transport systems have two equivalent "conserved components" acting either as the homologous halves of a single protein or as separate proteins (KERPPOLA et al. 1991, AMES and JOSHI 1990). Therefore, the N-terminal domain of NrtC can provide a second ATP-binding site. A function for the C-terminal domain of NrtC is currently unknown.

In summary, the genes that appear to be involved in nitrate transport in *Synechococcus* sp. PCC 7942 encode proteins which can constitute a transport system of the type known as "multicomponent", which use ATP as a source of energy for transport. Genes homologous to the nitrate transport genes of *Synechococcus* sp. are also present in *Anabaena* sp. PCC 7120 (FRÍAS et al. 1993).

5.3 The *narB* gene and other genes involved in nitrate reduction

The gene *narB* has been located in the cluster (LUQUE et al. 1992) just downstream from *nrtD* (OMATA et al. 1993). *narB* is one of three genes from *Synechococcus* sp. PCC 7942, namely *narA*, *narB*, and *narC*, that were identified as involved in nitrate reduction to nitrite and cloned by their ability to complement nitrate-reductase-deficient mutants, or by marker-rescue of transposon-induced nitrate-reductase-deficient mutants (KUHLEMEIER et al. 1984a, b). Nothing is currently known about the actual function of the products of either the *narA* or *narC* genes. The *narB* gene has been sequenced. Based on the similarity of the predicted amino acid sequence of the NarB protein to that of the α-subunit of *E. coli* nitrate reductase and other molybdo-proteins, and to the similarity of its amino acid composition to that of *Plectonema boryanum* nitrate reductase

(MIKAMI and IDA 1984), *narB* has been suggested to correspond to the nitrate reductase structural gene (ANDRIESSE et al. 1990). No evidence for the presence of either *narA* or *narC* in the nitrate-assimilation gene cluster has been obtained by means of either comparison of the restriction maps of the cloned genomic fragments containing those genes with that of the cluster, or by Southern hybridization between the genomic region containing the gene-cluster and the *narA* or *narC* genes (LUQUE et al. 1992). It seems therefore that both *narA* and *narC* are located somewhere else in the genome of *Synechococcus* sp. The *narA* and *narC* genes could be involved in the assembly of the molybdenum cofactor of nitrate reductase, their expression not being linked to that of the structural genes in the cluster.

With respect to the metabolism of molybdenum in cyanobacteria, mutant strains have been isolated from the filamentous N_2-fixer, *Anabaena variabilis* which are unable to synthesize active nitrate reductase, and one of them also nitrogenase, under normal, low-molybdenum growth conditions (MARTÍN-NIETO et al. 1990). These deficiencies were restorable by the addition of high levels of molybdate to the medium (MARTÍN-NIETO et al. 1990). From these results, the existence in *A. variabilis* of common genes for the biosynthesis of the molybdenum cofactors of nitrogenase and nitrate reductase, and specific genes for the biosynthesis of the molybdenum cofactor of nitrate reductase, was inferred (MARTÍN-NIETO et al. 1990).

On the other hand, the cloning of a DNA fragment from *Synechococcus* sp. PCC 6301 that would exhibit some homology to the *E. coli narG* gene, which encodes the α-subunit of respiratory nitrate reductase, has been reported (LIGHTFOOT et al. 1992). However, the actual function of that DNA fragment remains to be established.

5.4 Cotranscription of genes in the nitrate assimilation gene cluster

Mutant strains derived from *Synechococcus* sp. PCC 7942 have been described that are unable to grow at the expenses of nitrate and show a pleiotropic phenotype concerning nitrate assimilation. Strain FM16 (MADUEÑO et al. 1988a), that was isolated by transposon mutagenesis, exhibited low but appreciable nitrate reductase, high nitrite reductase, and non-detectable nitrate transport activity. The low nitrate reductase activity exhibited by this strain is, in contrast to what is the case in the wild-type strain, constitutive, i.e., not subject to ammonium-promoted repression (see below). Strain FM2 (MADUEÑO et al. 1988a) has Tn*901* inserted in the nitrite reductase gene (LUQUE et al. 1992a) and, as the result of this insertion, it lacks nitrite reductase activity and constitutively expresses nitrate transport and nitrate reductase activities. Finally, insertional mutants which were generated with a Km^r-encoding gene cassette inserted at defined positions inside the nitrite reductase gene (strains CSI3 through CSI6, LUQUE et al. 1992) exhibit undetectable levels of nitrite reductase; in all of these mutants, irrespective of the orientation of the inserted cassette, nitrate reductase activity was severely impaired. These polar effects of different transposon or gene-cassette insertions together with the physical linkage of the genes encoding nitrite reductase, the nitrate transport proteins, and presumptively nitrate reductase, indicate that the nitrate assimilation genes in this cluster are cotranscribed (LUQUE et al. 1992).

By means of Northern experiments, the presence in wild-type cells of *Synechococcus* sp. PCC 7942 of a transcript of ca. 1.4 kb, corresponding to that expected for a mRNA encoding only the NrtA protein, has been described (OMATA et al. 1989); however, this should be taken with caution in view of the above data indicating that the bulk of expression of the genes in the *Synechococcus* nitrate assimilation gene cluster takes place through cotranscription. Actually, more recently, different probes from the nitrate assimilation cluster have been reported to hybridize to degraded RNA of up to 7.5 kb which is only present after derepression of the nitrate assimilation system (SUZUKI et al. 1992). These results agree with the behaviour of the nitrate assimilation cluster as an operon.

The structure of the nitrate-assimilation gene-cluster of *Synechococcus* sp. PCC 7942 is summarized in Figure 2.

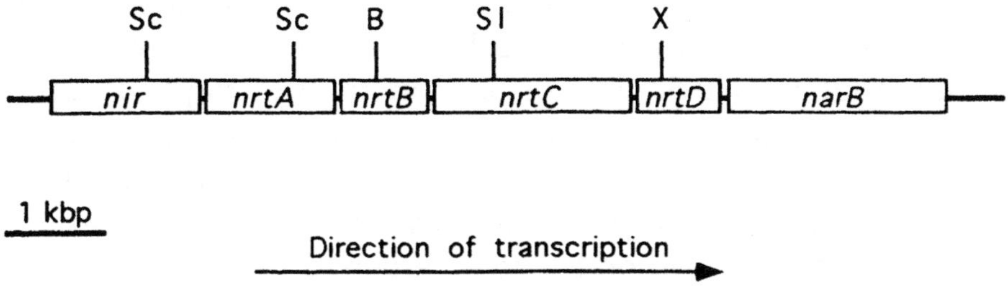

Fig. 2. Nitrate-assimilation gene-cluster found in the genome of *Synechococcus* sp. PCC 7942. The *nir* gene encodes nitrite reductase and the *nrt* genes encode a multicomponent nitrate transporter, whereas *narB* is the putative structural gene for nitrate reductase. Some restriction sites (B, *Bam*HI; Sc, *Sac*I; Sl, *Sal*I; X, *Xba*I) where antibiotic-resistance gene-cassettes were inserted are shown. Everyone of these insertions provoked polar effects on the expression of the genes located downstream of the insertion. The figure is redrawn from data of LUQUE et al. (1992, 1993) and OMATA et al. (1989, 1993).

6 Regulation of nitrate assimilation

Cyanobacteria preferentially use inorganic nitrogen as nitrogen source, namely, nitrate, nitrite, ammonium, and atmospheric nitrogen. In these organisms, a hierarchy has been shown to exist in the order in which these compounds are preferred to be assimilated. Thus, when ammonium is present in the external medium the system for nitrate assimilation is expressed only at low basal levels. The inhibition of nitrate assimilation exerted by ammonium operates at least at two different levels. In the short-term, it is manifest as an almost complete cessation of the utilization of nitrate immediately after the addition of ammonium to a suspension of cells actively taken up nitrate. If the cells continue to be exposed to ammonium, even when nitrate is simultaneously present, a long-term regulatory system operates which results in repression of the proteins involved in nitrate assimilation.

6.1 Regulation of the activity of the nitrate transport system

When cells of *Anabaena* sp., *Phormidium laminosum*, or *Synechococcus* sp. actively taken up nitrate are exposed to ammonium, a drastic inhibition of nitrate uptake takes place (OHMORI et al. 1977, FLORES et al. 1980, RAI et al. 1981, GARBISU et al. 1992) which operates at the level of nitrate transport (FLORES et al. 1980, LARA et al. 1987). This effect of ammonium on nitrate utilization is reversible. Thus, if low concentrations (10-50 µM) of ammonium are supplied, nitrate utilization resumes some minutes later, once ammonium has been exhausted from the medium (OHMORI et al. 1977, FLORES et al. 1980). The inhibition of nitrate uptake by ammonium requires that ammonium be incorporated into carbon skeletons through the operation of the glutamine synthetase/glutamate synthase pathway, since no negative effect of ammonium on nitrate utilization is observed if glutamine synthetase or glutamate synthase is inactivated, e.g., by treatment of the cells with MSX (FLORES et al. 1980, LARA et al. 1987) or azaserine (FLORES et al. 1980), respectively.

For *Anabaena cylindrica* and *Nostoc muscorum*, the proposition has been made that ammonium may inhibit nitrate uptake by reducing the supply of ATP which is required for active nitrate transport (OHMORI et al. 1977, RAI et al. 1981, RAI and SINGH 1982), as metabolism of ammonium through glutamine synthetase consumes ATP. However, only a transient decrease in the ATP level was observed to take place in cells of *A. cylindrica* in response to ammonium addition, the original high levels being soon recovered even when inhibitory concentrations of ammonium were still present in the medium (OHMORI and HATTORI 1978). More likely, it seems that an organic product formed through the assimilation of ammonium via glutamine synthetase could have a role in the regulation of nitrate uptake by acting as a feed-back inhibitor of this process (FLORES et al. 1980). Although MSX-treated cells of *Synechococcus* sp. PCC 6301, which are unable to manifest ammonium-promoted inhibition, exhibit lower levels of glutamine than the untreated controls (FLORES et al. 1983b) and glutamine inhibits nitrate uptake when added to the external medium (ROMERO et al. 1985a), some other experimental data speak against glutamine as the only metabolite regulating nitrate transport. Thus, in azaserine-treated cells, in which ammonium is also unable to inhibit nitrate transport, high levels of glutamine can be measured to accumulate (FLORES et al. 1983b). Also, intracellular levels of no amino acid significantly differs when cells of *Synechococcus* sp. PCC 6301 are steadily assimilating either nitrate or ammonium (CORONIL and LARA 1991). Therefore, it has been suggested that rather than by glutamine itself, ammonium inhibition could be mediated by some nitrogenous metabolite(s) generated in reactions which involve glutamine as the amido N donor (ROMERO et al. 1985a).

Nitrate uptake by *Synechococcus* sp. PCC 6301 shows a strict dependence upon CO_2 fixation which in the case of nitrogen-starved cells can be partially relieved by mobilization of accumulated carbohydrates (FLORES et al. 1983c). Two lines of evidence support the proposal that, in fact, nitrate uptake requires not only the presence of CO_2 but also formation of CO_2-fixation products. Thus, (i) the concentration dependence for CO_2 of nitrate uptake closely matches that of photosynthetic CO_2-fixation (FLORES et al. 1983c), and (ii) inhibition of photosynthetic CO_2-fixation by treatment of the cells with D,L-glyceraldehyde results in the simultaneous inhibition of nitrate utilization (ROMERO et al. 1985b). The requirement for CO_2-fixation products exhibited by nitrate transport suggests that these products may have a regulatory, positive action on the process, their role on nitrate assimilation being not merely to provide the carbon skeletons to which the ammonium resulting from nitrate reduction should be incorporated. If the requirement for CO_2-fixation products were only located at the level of ammonium assimilation, nitrate would be expected to be taken up and reduced by the cells, the resulting ammonium being excreted to the outer medium, as it is the case when ammonium assimilation is inhibited by treatment with MSX.

In MSX-treated cells of *Synechococcus* sp. PCC 6301 that cannot assimilate ammonium (FLORES et al. 1983c, ROMERO et al. 1985b), or in a nitrite reductase mutant of *Synechococcus* sp. PCC 7002 in which nitrate is taken up and reduced to nitrite that is excreted by the cells (STEVENS and VAN BAALEN 1972), nitrate is practically freed from its requirement for CO_2-fixation products. Also, CO_2 has a positive effect on the MSX-promoted restoration of nitrate uptake after its inhibition by ammonium (FLORES et al. 1983c). These facts establish a link between the two regulatory factors controlling nitrate uptake in *Synechococcus* sp., namely the negative effect of ammonium assimilation and the positive effect of CO_2-fixation, the two effects having antagonic roles on the regulation of nitrate uptake. The proposal has been made (FLORES et al. 1983c) that the CO_2-fixation products could combine with the negative effectors derived from ammonium assimilation, removing these latter compounds. On the other hand, it has been shown that generally a positive relationship exists between the susceptibility of a given amino acid to enzymically transfer its α-amino group to α-ketoglutarate and its ability to inhibit nitrate utilization (ROMERO et al. 1985). The inhibitory effect of some amino acids could therefore result from removal of α-ketoglutarate, which would play a positive role in the regulation of nitrate uptake.

In summary, nitrate uptake by *Synechococcus* cells seems to be controlled in an interactive manner by products of ammonium assimilation (negative) and CO_2-fixation (positive). Although the actual effector(s) ultimately interacting with the nitrate transport proteins remain to be identified, it can be envisaged that there could be both positive and negative direct effectors, or only positive or negative with the ammonium-derived or CO_2-derived products, respectively, contributing to their removal.

Nitrite uptake in *Synechococcus* sp. PCC 6301 (FLORES et al. 1987) and heterocystous cyanobacteria (MARTÍN-NIETO et al. 1989) has been described to take place both passively and actively (see above). The active transport of nitrite is, as it is also the case for nitrate transport, inhibited by products of the assimilation of ammonium via glutamine synthetase (FLORES et al. 1987). This could rather be expected from the above discussed evidence pointing to a common system for the operation of nitrate transport and nitrite active transport, which could thus be expected to be regulated by the same mechanism.

6.2 Regulation of the expression of proteins involved in nitrate assimilation

It has long been known that when cyanobacteria are using nitrate as the sole nitrogen source, they exhibit higher activity levels of nitrate reductase than when are exposed to ammonium. When cells of a variety of cyanobacteria are transferred from ammonium-containing media to media containing nitrate (or nitrite) as the sole nitrogen source, a striking increase in nitrate reductase activity takes place (STEVENS and VAN BAALEN 1974, HERRERO et al. 1981, HERRERO et al. 1985, MARTÍN-NIETO et al. 1989, PALOD et al. 1990). This increase in cellular levels is also observed for nitrite reductase activity (OHMORI and HATTORI 1970, MÉNDEZ et al. 1981, HERRERO and GUERRERO 1986, ARIZMENDI et al. 1987, MARTÍN-NIETO et al. 1989, PALOD et al. 1990) as well as for the 48-kDa cytoplasmic-membrane protein involved in nitrate transport (MADUEÑO et al. 1988b, OMATA et al. 1989, SIVAK et al. 1989).

The degree of responsiveness of the nitrate reduction system to the nature of the available nitrogen source varies in magnitude among different strains of cyanobacteria. Control seems to be tight in *Synechococcus* sp. PCC 6301, *Anabaena* sp. ATCC 33047, *A. variabilis* ATCC 29413, *Anabaena* sp. PCC 7120, *Nostoc* sp. ATCC 29150, *Calothrix* sp. PCC 7101, and *Calothrix* sp. PCC 7601, whereas it is weaker in *Synechococcus* sp. PCC 7942, *Synechocystis* sp. PCC 6714, *A. cylindrica* ATCC 29414, *Anabaena* sp. PCC 7119, and *Nostoc* sp. PCC 6719 (MEEKS et al. 1983, RAMOS and GUERRERO 1983, HERRERO et al. 1985, HERRERO and GUERRERO 1986, MADUEÑO et al. 1988a, MARTÍN-NIETO et al. 1989).

For both nitrate reductase and nitrite reductase, the increase in activity upon transference to media lacking ammonium is fully inhibited by the presence of rifamycin, chloramphenicol, or erythromycin, from what an involvement of *de novo* protein synthesis was invoked (STEVENS and VAN BAALEN 1974, HATTORI 1962, HERRERO et al. 1981, HERRERO et al. 1985, HERRERO and GUERRERO 1986, ARIZMENDI et al. 1987, PALOD et al. 1990). From our own data this holds true also for the filamentous cyanobacterium *Anabaena variabilis* ATCC 29413 (HERRERO et al. 1985), though lack of chloramphenicol effect on the early development of nitrate reductase activity upon transference from ammonium-containing media to media containing nitrate has been reported for this strain (AVISSAR 1985).

Different responses to the supply of nitrate or nitrite have been found between the non-nitrogen-fixing and the filamentous, nitrogen-fixing cyanobacteria. While the non-nitrogen-fixers are able to develop high levels of both nitrate reductase and nitrite reductase in media lacking combined nitrogen (HERRERO et al. 1981, HERRERO et al. 1985, HERRERO and GUERRERO 1986, ARIZMENDI et al. 1987, PALOD et al. 1990), the presence of either nitrate or nitrite as inducers is necessary for activity development in the nitrogen fixers (HATTORI 1962, HERRERO et al. 1981, HERRERO et al. 1985, MARTÍN-NIETO et al. 1989). Arginine has been postulated to behave as an inducer of nitrate reductase in *Oscillatoria chalybea* and *Anabaena* sp. PCC 7120 (BEDNARZ and SCHMID 1991, 1992); however, according to our own data (HERRERO and FLORES 1990), arginine is not an inducer of nitrate reductase in *Anabaena* sp. PCC 7120.

In those instances where induction is necessary, ammonium has been established (MARTÍN-NIETO et al. 1989) to promote a true repression of the enzymes, its negative effect on enzyme synthesis not merely being exerted through inducer exclusion (as nitrate transport and active nitrite transport are inhibited by ammonium, see above). This postulate is based on the observation that under conditions in which nitrite majoritarily enters the cyanobacterial cell in a passive way (non-inhibitable by ammonium), nitrate reductase and nitrite reductase do not develop in a medium containing both nitrite and ammonium.

As it was stated above in the case of the ammonium-promoted regulation of nitrate transport, ammonium-promoted repression of nitrate reductase and nitrite reductase in cyanobacteria requires that ammonium be incorporated into organic material through the operation of glutamine synthetase. This is derived from the fact that ammonium is unable to exert any negative effect on nitrate reductase and nitrite reductase synthesis in cells that have been treated with MSX and therefore bear an inactive glutamine synthetase (HERRERO et al. 1981, RAMOS and GUERRERO 1983, HERRERO et al. 1985, HERRERO and GUERRERO 1986).

Stability of preformed nitrate reductase in *Synechococcus* sp. PCC 6301 is higher in the presence of nitrate or ammonium than in the absence of combined nitrogen. Thus, although nitrate seems not to be required as an inducer in this cyanobacterium, stabilization of the enzyme by nitrate could contribute to achieve the nitrate reductase activity levels found in the presence of nitrate that are higher than those found in the absence of combined nitrogen (HERRERO et al. 1984).

7 Nitrogen control in cyanobacteria

As mentioned above, both the activity of the nitrate/nitrite transport system and the synthesis of nitrate reductase and nitrite reductase in cyanobacteria are regulated through systems that, physiologically, share a common characteristic. In both processes, ammonium acts as a negative effector that exerts its action after its incorporation into carbon skeletons through the glutamine synthetase/glutamate synthase pathway. This fact spoke in favour of a common network regulating at more than one point nitrate assimilation in cyanobacteria. Other processes concerning nitrogen assimilation in cyanobacteria also exhibit similar features, as it is the case for nitrogenase synthesis and heterocyst development in filamentous, heterocyst-forming cyanobacteria (STEWART and ROWELL 1975). Thus, a general system operating global nitrogen control could be envisaged to exist in these organisms.

Mutant strains from the cyanobacterium *Synechococcus* sp. PCC 7942 were isolated (VEGA-PALAS et al. 1990) that, as the result of single mutations, exhibited a pleiotropic impairment in the expression of a number of activities/proteins involved in nitrogen assimilation, all of which are regulated through ammonium-promoted repression in the wild-type strain. These mutants exhibited low levels of the 48-kDa protein that is part of the nitrate transport system, nitrate reductase, nitrite reductase, glutamine synthetase and methylammonium transport (VEGA-PALAS et al. 1990). In every case, the levels of these activities/proteins in the pleiotropic mutants were, irrespective of the nitrogen content of the medium, similar to those found in ammonium-grown cells of the wild-type strain.

The mutated gene responsible for the pleiotropic phenotype of the mutants has a positive action on the expression of the wild-type proteins, since revertants could be isolated that always showed wild-type levels and pattern of regulation of all the mutated functions (VEGA-PALAS 1990). This gene was named *ntcA (ntc* standing for nitrogen control). The *ntcA* gene was cloned by complementation of one of the pleiotropic mutants and sequenced (VEGA-PALAS et al. 1992). Its predicted protein product (NtcA) was found to show homology to the proteins that are members of the family of bacterial transcriptional activators to which Crp (the cyclic-AMP receptor protein) and Fnr (that allows the use of electron acceptors other than oxygen under anaerobic conditions) from *E. coli* also belong.

As it is also the case for the other members of the family, NtcA bears a helix-turn-helix motif in its carboxy-terminal region through which it could interact with the promoter region of the regulated genes in order to operate activation of their transcription. In this context, it is worth mentioning that mobility-shift experiments that are being performed in the authors' laboratory (LUQUE et al. 1994) show that NtcA is able to bind to the promoter regions of both the *nir* and the glnA (encoding glutamine synthetase) genes.

The question of the presence in the cyanobacteria of a system homologous to the enterobacterial *ntr* system is of interest. The only component of the *ntr* system whose presence in the cyanobacteria is well documented is the P_{II} protein which in *E. coli* is involved in signal transduction to the Ntr proteins, as well as in the glutamine synthetase adenylylation cascade (MAGASANIK and NEIDHARDT 1987). The P_{II} protein was identified in *Synechococcus* sp. PCC 6301 as a soluble protein that is phosphorylated during growth under light preferentially absorbed by Photosystem II (HARRISON et al. 1990). The *glnB* gene encoding P_{II} has been cloned and sequenced from *Synechococcus* sp. PCC 7942 (TSINOREMAS et al. 1991). Identity of the 112-amino acid *Synechococcus* P_{II} (or GlnB) protein to the P_{II} proteins from five other bacteria is more than 60% and includes the tyrosine residue that is subject to uridylylation in *E. coli*. Phosphorylation of the *Synechococcus* P_{II} protein is negatively affected by ammonium and, therefore, it could be involved in signaling the nitrogen status of the cell.

8 Concluding remarks

Like in many other fields of biology, the study of nitrate assimilation by cyanobacteria has received a great impulse in the last few years with the application of the molecular-genetical approach. This (1) has led to the unraveling of the organization of some of the genes involved in nitrate assimilation; (2) has provided insight into the structure of nitrite reductase, which has revealed homologous to the enzyme from the chloroplast of higher plants, thus supporting the postulate of the chloroplast deriving from a cyanobacterium-like endosymbiont; and (3) has provided valuable information regarding the mechanism by which the classical subject of ammonium repression of nitrate assimilation operates at the molecular level. In not too long a time, molecular biology is expected to yield insight into the mechanism by which control of the activity of nitrate transport is operated, as well as more details on the mechanism of regulation of the expression of proteins involved in nitrate assimilation. In this context, it would be of special interest the investigation of how the action of the regulatory protein NtcA (which controls the expression of nitrate-assimilation proteins) is itself regulated in the cell in response to the nitrogen content of the environment, either by regulation of NtcA synthesis or/and by posttranslational modulation of its activity.

Acknowledgements

Work in the authors' laboratory is currently supported by grants from DGICYT (grant no. BP90-0114) and Junta de Andalucía (research group no. 3057), Spain. We thank T. OMATA for making available manuscripts before publication.

References

Ames GF-L (1986) Bacterial periplasmic transport systems: structure, mechanism, and evolution. Annu Rev Biochem 55:397-425

Ames GF-L, Joshi AK (1990) Energy coupling in bacterial periplasmic permeases. J Bacteriol 172:4133-4137

Andriesse X, van Arkel G, Weisbeek P (1990) Cloning of the nitrate reductase gene of *Synechococcus* PCC 7942. Third International Symposium on NO_3^- Assimilation: Molecular and Genetic Aspects, Bombanes, p 98

Arizmendi JM, Serra JL (1990) Purification and some properties of the nitrite reductase from the cyanobacterium *Phormidium laminosum*. Biochim Biophys Acta 1040:237-244

Arizmendi JM, Fresnedo O, Martínez-Bilbao M, Alaña A, Serra JL (1987) Inorganic nitrogen assimilation in the non-N_2-fixing cyanobacterium *Phormidium laminosum*. II. Effect of the nitrogen source on the nitrite reductase levels. Physiol Plant 70:703-707

Avissar YL (1985) Induction of nitrate assimilation in the cyanobacterium *Anabaena variabilis*. Physiol Plant 63:105-108

Bagchi SN, Sherman TD, Funkhouser EA (1987) Biochemical characterization of molybdenum-cofactor in a cyanobacterium, *Nostoc muscorum*. Plant Cell Physiol 28:1411-1419

Bagchi SN, Palod A, Chauhan VS (1989) Photosynthetic control of nitrate metabolism in *Phormidium uncinatum*, a cyanobacterium. Curr Microbiol 19:183-188

Bednarz J, Schmid GH (1991) Induction of nitrate reductase activity by arginine in the filamentous cyanobacterium *Oscillatoria chalybea*. Z. Naturforsch 46c:591-596

Bednarz J, Schmid GH (1992) Further studies on the induction of nitrate reductase by arginine in the filamentous cyanobacterium *Oscillatoria chalybea*. Z. Naturforsch 47c:540-544

Benz R, Böhme H (1985) Pore formation by an outer membrane protein of the cyanobacterium *Anabaena variabilis*. Biochim Biophys Acta 812:286-292

Bishop L, Agbayani R, Ambudkar SV, Maloney PC, Ames GF-L (1989) Reconstitution of a bacterial periplasmic permease in proteoliposomes and demonstration of ATP hydrolysis concomitant with transport. Proc Natl Acad Sci USA 86:6953-6957

Blanco F, Serra JL (1992) Purificación y algunas propiedades de la ferredoxina-nitrato reductasa de cultivos de *Phormidium laminosum* sometidos deficiencia en nitrógeno. In: Pineda M, Castillo F (eds) Metabolismo del nitrógeno, Sociedad Española de Bioquímica, Córdoba, pp 68-77

Candau P (1979) Purificación y propiedades de la ferredoxina-nitrato reductasa de la cianobacteria *Anacystis nidulans*. PhD Thes, Univ Sevilla, Sevilla

Candau P, Manzano C, Losada M (1976) Bioconversion of light energy into chemical energy through reduction with water of nitrate to ammonia. Nature 262:715-717

Candau P, Manzano C, Guerrero MG, Losada M (1980) Ferredoxin-dependent enzymatic reduction of nitrate with a deazaflavin photosystem. Photobiochem Photobiophys 1:167-174

Coronil T, Lara C (1991) Amino acid levels and expression of the nitrate assimilation system in *Anacystis nidulans* cells. Plant Physiol Biochem 29:651-655

Drews G, Weckesser J (1982) Function, structure and composition of cell walls and external layers. In: Carr NG, Whitton BA (eds) The biology of cyanobacteria, Blackwell Sci Pub, Oxford, pp 333-357

Flores E, Guerrero MG, Losada M (1980) Short-term ammonium inhibition of nitrate utilization by *Anacystis nidulans* and other cyanobacteria. Arch Microbiol 128:137-144

Flores E, Guerrero MG, Losada M (1983a) Photosynthetic nature of nitrate uptake and reduction in the cyanobacterium *Anacystis nidulans*. Biochim Biophys Acta 722:408-416

Flores E, Ramos JL, Herrero A, Guerrero MG (1983b) Nitrate assimilation by cyanobacteria. In: Papageorgiou GC, Packer L (eds) Photosynthetic prokaryotes: Cell differentiation and function, Elsevier Sci Pub Co Inc, New York, pp 363-387

Flores E, Romero JM, Guerrero MG, Losada M (1983c) Regulatory interaction of photosynthetic nitrate utilization and carbon dioxide fixation in the cyanobacterium *Anacystis nidulans*. Biochim Biophys Acta 725:529-532

Flores E, Herrero A, Guerrero MG (1987) Nitrite uptake and its regulation in the cyanobacterium *Anacystis nidulans*. Biochim Biophys Acta 896:103-108

Fogg GE (ed) (1975) Algal cultures and phytoplankton ecology. The University of Wisconsin Press Ltd, Madison

Frías JE, Mérida A, Herrero A, Martín-Nieto J, Flores E (1993) General distribution of the nitrogen control gene *ntcA* in cyanobacteria. J Bacteriol 175:5710-5713

Garbisu C, Hall DO, Serra JL (1992) Nitrate and nitrite uptake by free-living and immobilized N-starved cells of *Phormidium laminosum*. J Appl Phycol 4:139-148

García-Pichel F, Castenholz RW (1990) Comparative anoxygenic photosynthetic capacity in 7 strains of a thermophilic cyanobacterium. Arch Microbiol 153:344-351

Guerrero MG, Lara C (1987) Assimilation of inorganic nitrogen. In: Fay P, van Baalen C (eds) The cyanobacteria, Elsevier Sci Pub BV, Amsterdam, pp 163-186

Guerrero MG, Manzano C, Losada M (1974) Nitrite photoreduction by a cell-free preparation of *Anacystis nidulans*. Plant Sci Lett 3:273-278

Harrison MA, Keen JN, Findlay JBC, Allen JF (1990) Modification of a glnB-like gene product by photosynthetic electron transport in the cyanobacterium *Synechococcus* 6301. FEBS Lett 264:25-28

Hattori A (1962) Adaptive formation of nitrate reducing system in *Anabaena cylindrica*. Plant Cell Physiol 3:371-377

Hattori A, Myers J (1966) Reduction of nitrate and nitrite by subcellular preparations of *Anabaena cylindrica*. I. Reduction of nitrite to ammonia. Plant Physiol 41:1031-1036

Hattori A, Myers J (1967) Reduction of nitrate and nitrite by subcellular preparations of *Anabaena cylindrica*. II. Reduction of nitrate to nitrite. Plant Cell Physiol 8:327-337

Hattori A, Uesugi I (1968) Purification and properties of nitrite reductase from the blue-green alga *Anabaena cylindrica*. Plant Cell Physiol 9:689-699

Herrero A, Flores E (1990) Transport of basic amino acids by the dinitrogen-fixing cyanobacterium *Anabaena* PCC 7120. J Biol Chem 265:3931-3935

Herrero A, Guerrero MG (1986) Regulation of nitrite reductase in the cyanobacterium *Anacystis nidulans*. J Gen Microbiol 132:2463-2468

Herrero A, Flores E, Guerrero MG (1981) Regulation of nitrate reductase levels in the cyanobacteria *Anacystis nidulans, Anabaena* sp. strain 7119, and *Nostoc* sp. strain 6719. J Bacteriol 145:175-180

Herrero A, Flores E, Guerrero MG (1984) Regulation of the nitrate reductase level in *Anacystis nidulans*: activity decay under nitrogen stress. Arch Biochem Biophys 234:454-459

Herrero A, Flores E, Guerrero MG (1985) Regulation of nitrate reductase cellular levels in the cyanobacteria *Anabaena variabilis* and *Synechocystis* sp. FEMS Microbiol Lett 26:21-25

Ida S, Mikami B (1983) Purification and characterization of assimilatory nitrate reductase from the cyanobacterium *Plectonema boryanum*. Plant Cell Physiol 24:649-658

Kerppola RE, Shyamala VK, Klebba P, Ames GF-L (1991) The membrane-bound proteins of periplasmic permeases form a complex. Identification of the histidine permease HisQMP complex. J Biol Chem 266:9857-9865

Kuhlemeier CJ, Logtenberg T, Stoorvogel W, van Heugten HAA, Borrias WE, van Arkel GA (1984a) Cloning of nitrate reductase genes from the cyanobacterium *Anacystis nidulans*. J Bacteriol 159:36-41

Kuhlemeier CJ. Teeuwsen VJP, Janssen MJT, van Arkel GA (1984b) Cloning of a third nitrate reductase gene from the cyanobacterium *Anacystis nidulans* R2 using a shuttle gene library. Gene 31:109-116

Lara C, Romero JM, Guerrero MG (1987) Regulated nitrate transport in the cyanobacterium *Anacystis nidulans*. J Bacteriol 169:4376-4378

Lightfoot DA, Baron AJ, Cock JM, Wootton JC (1992) A nitrate reductase gene of the cyanobacterium *Synechococcus* PCC 6301 inferred by heterologous hybridization, cloning and targeted mutagenesis. Genetica 85:107-117

Luque I, Herrero A, Flores E, Madueño F (1992) Clustering of genes involved in nitrate assimilation in the cyanobacterium *Synechococcus*. Mol Gen Genet 232:7-11

Luque I, Flores E, Herrero A (1993) Nitrite reductase gene from *Synechococcus* sp. PCC 7942: homology between cyanobacterial and higher-plant nitrite reductases. Plant Mol Biol 22:1201-1205

Luque I, Flores E, Herrero A (1994) Molecular mechanism for the operation of nitrogen control in cyanobacteria. Embo J 13:2862-2869

Madueño F, Flores E, Guerrero MG (1987) Competition between nitrate and nitrite uptake in the cyanobacterium *Anacystis nidulans*. Biochim Biophys Acta 896: 109-112

Madueño F, Borrias WE, van Arkel GA, Guerrero MG (1988a) Isolation and characterization of *Anacystis nidulans* R2 mutants affected in nitrate assimilation: establishment of two new mutant types. Mol Gen Genet 213:223-228

Madueño F, Vega-Palas MA, Flores E, Herrero A (1988b) A cytoplasmic-membrane protein repressible by ammonium in *Synechococcus* R2: altered expression in nitrate-assimilation mutants. FEBS Lett 239:289-291

Magasanik B, Neidhardt FC (1987) Regulation of carbon and nitrogen utilization. In: Ingraham JL, Low KB, Magasanik B, Schaechter M, Umbarger HE (eds) *Escherichia coli* and *Salmonella typhimurium*. Cellular and molecular biology, American Society for Microbiology, Washington, pp 1318-1325

Maloney PC (1990) Resolution and reconstitution of anion exchange. Philos Trans R Soc Lond B Biol Sci 326:437-454

Manzano C (1977) La reducción fotosintética del nitrato en el alga verde-azulada *Anacystis nidulans*. Ph D Thes, Univ Sevilla, Sevilla

Manzano C, Candau P, Gómez-Moreno C, Relimpio AM, Losada M (1976) Ferredoxin-dependent photosynthetic reduction of nitrate and nitrite by particles of *Anacystis nidulans*. Mol Cell Biochem 10:161-169

Martín-Nieto J, Herrero A, Flores E (1989) Regulation of nitrate and nitrite reductases in dinitrogen-fixing cyanobacteria and Nif^- mutants. Arch Microbiol 151:475-478

Martín-Nieto J, Flores E, Herrero A (1990) Mutants of *Anabaena variabilis* requiring high levels of molybdate for nitrate reductase and nitrogenase activities. FEMS Microbiol Lett 67:1-4

Martín-Nieto J, Flores E, Herrero A (1992) Biphasic kinetic behavior of nitrate reductase from heterocystous, nitrogen-fixing cyanobacteria. Plant Physiol 100:157-163

Meeks JC, Wolk CP, Thomas J, Lockau W, Shaffer PW, Austin SM, Chien W-S, Galonsky A (1977) The pathways of assimilation of $^{13}NH_4^+$ by the cyanobacterium, *Anabaena cylindrica*. J Biol Chem 252:7894-7900

Meeks JC, Wycoff KL, Chapman JS, Enderlin CS (1983) Regulation of expression of nitrate and dinitrogen assimilation of $[^{13}N]N_2$ and $^{13}NH_4^+$ by cyanobacteria with and without heterocysts. J Bacteriol 134:125-130

Meeks JC, Wycoff KL, Chapman JS, Enderlin CS (1983) Regulation of expression of nitrate and dinitrogen assimilation by *Anabaena* species. Appl Environm Microbiol 45:1351- 1359

Méndez JM, Vega JM (1981) Purification and molecular properties of nitrite reductase from *Anabaena* sp. 7119. Physiol Plant 52:7-14

Méndez JM, Herrero A, Vega JM (1981) Characterization and catalytic properties of nitrite reductase from *Anabaena* sp. 7119. Z Pflanzenphysiol 103:305-315

Mikami B, Ida S (1984) Purification and properties of ferredoxin-nitrate reductase from the cyanobacterium *Plectonema boryanum*. Biochim Biophys Acta 791:294-304

Mikami B, Ida S (1986) Reversible inactivation of ferredoxin-nitrate reductase from the cyanobacterium *Plectonema boryanum*. The role of superoxide anion and cyanide. Plant Cell Physiol 27: 1013-1021

Ohmori K, Hattori A (1970) Induction of nitrate and nitrite reductases in *Anabaena cylindrica*. Plant Cell Physiol 11:873- 878

Ohmori M, Hattori A (1978) Transient change in the ATP pool of *Anabaena cylindrica* associated with ammonia assimilation. Arch Microbiol 117:17-20

Ohmori M, Ohmori K, Strotmann H (1977) Inhibition of nitrate uptake by ammonia in a blue-green alga, *Anabaena cylindrica*. Arch Microbiol 114:225-229

mata T (1991) Cloning and characterization of the *nrtA* gene that encodes a 45-kDa protein involved in nitrate transport in the cyanobacterium *Synechococcus* PCC 7942. Plant Cell Physiol 32:151-157

Omata T, Ohmori M, Arai N, Ogawa T (1989) Genetically engineered mutant of the cyanobacterium *Synechococcus* PCC 7942 defective in nitrate transport. Proc Natl Acad Sci USA 86:6612- 6616

Omata T, Andriesse X, Hirano A (1993) Identification and characterization of a gene cluster involved in nitrate transport of the cyanobacterium *Synechococcus* sp. PCC 7942. Mol Gen Genet 236:193-202

Ortega T, Castillo F, Cárdenas J (1976) Photolysis of water coupled to nitrate reduction by *Nostoc muscorum* subcellular particles. Biochem Biophys Res Commun 71:885-891

Osborn MJ, Wu HCP (1980) Proteins of the outer membrane of Gram-negative bacteria. Annu Rev Microbiol 34:369-422

Ostrowski J, Wu J-Y, Rueger DC, Miller BE, Siegel LM, Kredich NM (1989) Characterization of the *cysJIH* regions of *Salmonella typhimurium* and *Escherichia coli* B. J Biol Chem 264:15726-15737

Palod A, Chauhan VS, Bagchi SN (1990) Regulation of nitrate reduction in a cyanobacterium *Phormidium uncinatum*: Distinctive modes of ammonium-repression of nitrate and nitrite reductases. FEMS Microbiol Lett 68:285-288

Paschinger H (1977) DCCD induced sodium uptake by *Anacystis nidulans*. Arch Microbiol 113:285-291

Peschek GA (1979) Nitrate and nitrite reductase and hydrogenase in *Anacystis nidulans* grown in Fe- and Mo-deficient media. FEMS Microbiol Lett 6:371-374

Rai AK, Kashyap AK, Gupta SL (1981) ATP-dependent uptake of nitrate in *Nostoc muscorum* and inhibition by ammonium ions. Biochim Biophys Acta 674:78-86

Rai AK, Singh S (1982) Regulation of nitrate uptake in *Nostoc muscorum* by glutamine synthetase. FEMS Microbiol Lett 14:303-306

Ramos JL, Guerrero MG (1983) Involvement of ammonium metabolism in the nitrate inhibition of nitrogen fixation in *Anabaena* sp. strain ATCC 33047. Arch Microbiol 136:81-83

Ramos JL, Guerrero MG, Losada M (1982) Photoproduction of ammonia from nitrate by *Anacystis nidulans* cells. Biochim Biophys Acta 679:323-330

Rippka R, Waterbury JB, Stanier RY (1981) Isolation and purification of cyanobacteria: Some general principles. In: Starr MP, Stolp H, Trüer HG, Balows A, Schlegel HG (eds) The prokaryotes: A handbook on habitats, isolation, and identification of bacteria, vol I. Springer-Verlag, Berlin, pp 212-220

Rodríguez R, Lara C, Guerrero MG (1992) Nitrate transport in the cyanobacterium *Anacystis nidulans* R2. Kinetic and energetic aspects. Biochem J 282:639-643

Romero JM, Flores E, Guerrero MG (1985a) Inhibition of nitrate utilization by amino acids in intact *Anacystis nidulans* cells. Arch Microbiol 142:1-5

Romero JM, Lara C, Guerrero MG (1985b) Dependence of nitrate utilization upon active CO_2 fixation in *Anacystis nidulans*: A regulatory aspect of the interaction between photosynthetic carbon and nitrogen metabolism. Arch Biochem Biophys. 237:396-401

Serrano A, Losada M (1988) Action spectra for nitrate and nitrite assimilation in blue-green algae. Plant Physiol 86:1116- 1119

Serrano A, Rivas J, Losada M (1981) Nitrate and nitrite as 'in vivo' quenchers of chlorophyll fluorescence in blue-green algae. Photosynth. Res. 2:175-184

Serrano A, Rivas J, Losada M (1982) Changes in fluorescence spectra by nitrate and nitrite in a blue-green alga. Photobiochem Photobiophys 4:257-264

Shearer G, Schneider JD, Kohl DH (1991) Separating the efflux and influx components of net nitrate uptake by *Synechococcus* R2 under steady-state conditions. J Gen Microbiol 137:1179-1184

Sivak MN, Lara C, Romero JM, Rodríguez R, Guerrero MG (1989) Relationship between a 47-kDa cytoplasmic membrane polypeptide and nitrate transport in *Anacystis nidulans*. Biochem Biophys Res Commun 158:257-262

Stevens SE, van Baalen C (1973) Characteristics of nitrate reduction in a mutant of the blue-green alga *Agmenellum quadruplicatum*. Plant Physiol 51:350-356

Stevens SE, van Baalen C (1974) Control of nitrate reductase in a blue-green alga. The effects of inhibitors, blue light, and ammonia. Arch Biochem Biophys 161:146-152

Stewart WDP, Rowell P (1975) Effects of L-methionine-DL- sulphoximine on the assimilation of newly fixed NH_3, acetylene reduction and heterocyst production in *Anabaena cylindrica*. Biochem Biophys Res Commun 65:846-856

Suzuki I, Sugiyama T, Omata T (1992) Gene expression and regulation of nitrate assimilating enzymes in *Synechococcus* PCC 7942. In: Murata N (ed) Research in photosynthesis, vol 1. Kluwer Academic Publishers, Dordrecht, The Netherlands, pp 75-78

Syrett PJ (1981) Nitrogen metabolism of microalgae. Can Bull Fish Aquat Sci 210:182-210

Tischner R, Schmidt A (1984) Light mediated regulation of nitrate assimilation in *Synechococcus leopoliensis*. Arch Microbiol 137:151-154

Tsinoremas NF, Castets AM, Harrison MA, Allen JF, Tandeau de Marsac N (1991) Photosynthetic electron transport controls nitrogen assimilation in cyanobacteria by means of posttranslational modification of the *glnB* gene product. Proc Natl Acad Sci USA 88:4565-4569

Vega-Palas MA, Madueño F, Herrero A, Flores E (1990) Identification and cloning of a regulatory gene for nitrogen assimilation in the cyanobacterium *Synechococcus* sp. strain PCC 7942. J Bacteriol 172:643-647

Vega-Palas MA, Flores E, Herrero A (1992) NtcA, a global nitrogen regulator from the cyanobacterium *Synechococcus* that belongs to the Crp family of bacterial regulators. Mol Microbiol 6:1853-1859

Wolk CP, Thomas J, Shaffer PW, Austin SM, Galonsky A (1976) Pathway of nitrogen metabolism after fixation of ^{13}N-labeled nitrogen gas by the cyanobacterium, *Anabaena cylindrica*. J Biol Chem 251:5027-5034

Yabuki Y, Mori E, Tamura G (1985) Nitrite reductase in the cyanobacterium *Spirulina platensis*. Agric Biol Chem 49:3061-3062

Zevenboom W, Mur LR (1979) Influence of growth rate on short term and steady state nitrate uptake by nitrate-limited *Oscillatoria agardhii*. FEMS Microbiol Lett 6:209-212

Ammonia Assimilation and its Biotechnological Aspects in Cyanobacteria

S. BOUSSIBA

The Jacob Blaustein Institute for Desert Research, Microalgal Biotechnology, Ben-Gurion University of the Negev, Sede Boker Campus, Israel 84990

1 Introduction

Ammonia occupies an unique biochemical position, being the only inorganic form of nitrogen that is directly incorporated into organic linkage, and thus an obligate intermediate in the utilization of other inorganic nitrogen sources. Ammonia is the most preferable nitrogen source and when supplied together with nitrate, algae will generally deplete first the ammonia then utilize the other nitrogen source. This molecule is also well known as uncoupler in photosynthesis, causing under certain circumstances the collapse of ΔpH required for ATP synthesis.

Ammonia regulates also key enzyme in nitrogen metabolism such as nitrate reductase, glutamine synthetase and nitrogenase. Studying the pathway through which this molecule enters and exits cyanobacterial cells may elucidate the process through which this regulation occurs. Specific transport systems have evolved to deal with such problems, and over the last years, systems for concentrative uptake of ammonium ions have been defined in many free-living microorganisms (see for reviews: KLEINER 1981, 1985, BOUSSIBA and GIBSON 1991).

The biotechnological implications of this molecule, mainly in tropical agriculture involve its substantial production and release under certain circumstances. The fact that this molecule can act also as uncoupler of photosynthesis has been used also in other fields of algal biotechnology.

This chapter will discuss different topics related to this molecule; its route of assimilation in cyanobacteria, the mechanisms involved in its penetration and exit from these cells and finally with the biotechnological aspects related to this nitrogen molecule.

2 Ammonium assimilation

Ammonium taken up from the outer medium or internally generated by either nitrate reduction or dinitrogen fixation is assimilated through its combination with carbon skeletons. In parallel to this primary assimilation, the ammonium resulting from the photorespiratory conversion of glycine to serine (CODD and

STEWART 1973, BERGMAN 1984) or from protein degradation followed by deamination and deamidation of amino acids (BOUSSIBA et al. 1984), is also reassimilated.

2.1 Pathways

Available evidence indicates that for a wide range of cyanobacteria, the glutamine synthetase (GS)-glutamate synthase (GOGAT) enzyme system is the quantitatively most important pathway for the assimilation of ammonium, either exogenuously or internally generated. In this route (Fig. 1), ammonium-N is initially incorporated into the amide position of glutamine, from where it is transferred to the α-position of α-ketoglutarate. One ATP molecule and two reducing equivalents are required for the net formation of one glutamate unit from ammonium and α-ketoglutarate (STEWART 1980, WOLK 1980, MEEKS 1981, LEA et al. 1982, FLORES et al. 1983). Evidence supporting this includes: (a) ^{13}N (THOMAS et al. 1975, 1977, WOLK et al. 1976, MEEKS et al. 1977, 1978, 1985, MEEKS 1981) and ^{15}N (STEWART et al. 1975) tracer studies; (b) use of glutamate analogs, inhibitors of either enzyme of the pathway, such as MSX, azaserine, 5-hydroxylysine, 6-diazo-5-oxonorleucine, and phosphinotricine (STEWART and ROWELL 1975, WOLK et al. 1976, MEEKS et al. 1977, 1978, THOMAS et al. 1977, FLORES et al. 1980, 1983b, RAMOS et al. 1982, BERGMAN 1984, BOUSSIBA et al. 1984c, BOUSSIBA 1991, LEA et al. 1984); and (c) detection of the enzymes and evaluation of their cellular activity level (HAYSTEAD et al. 1973, STEWART et al. 1975, STEWART 1980b).

The net glutamate unit resulting from ammonium assimilation through the concerted participation of GS and GOGAT can donate its amino group to a variety of carbon skeletons in reactions catalyzed by transaminases. Some of the glutamine synthesized by the action of GS does not transfer its amide group to α-ketoglutarate but to various substrates of glutamine-amide transferases other than GOGAT (Fig. 1).

In addition to the GS-GOGAT system, other enzymes like glutamate dehydrogenase and alanine dehydrogenase may be active in ammonium assimilation in cyanobacteria (MEEKS et al. 1977,

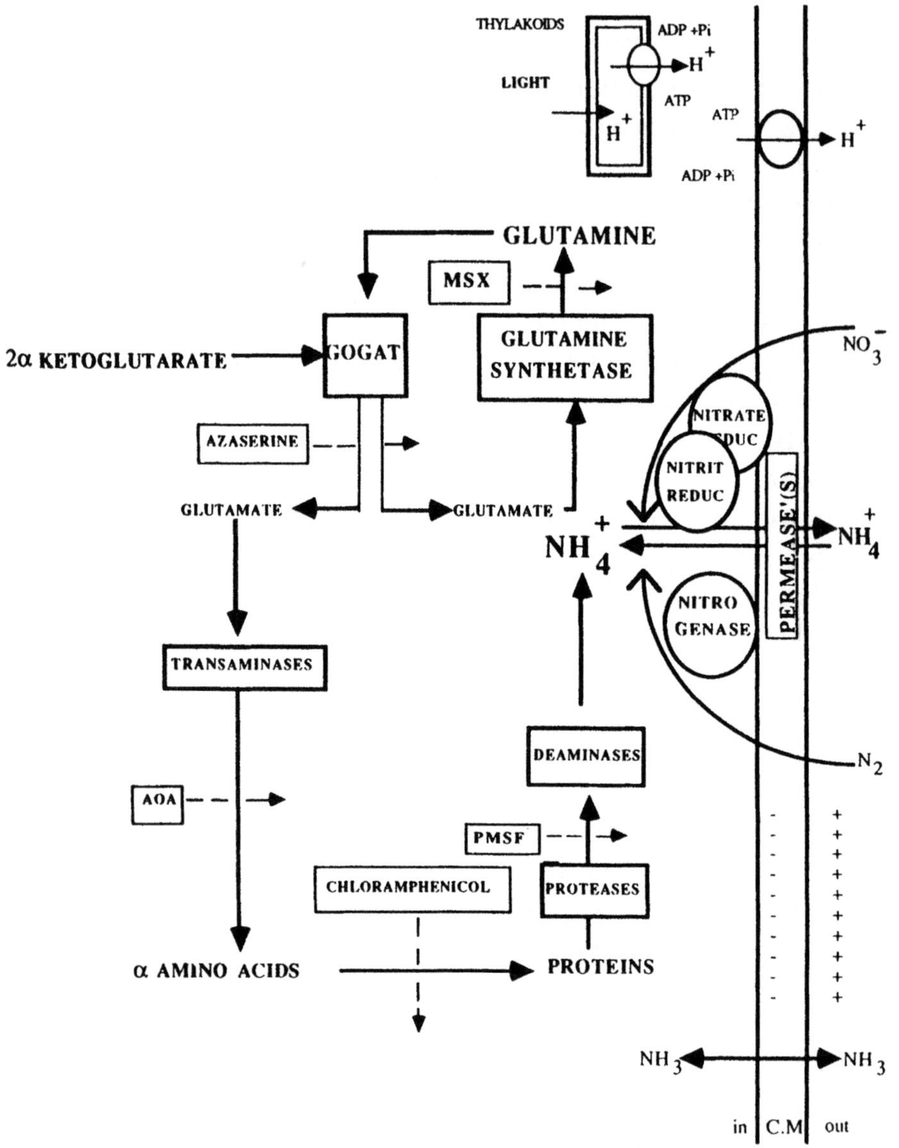

Fig. 1. Ammonium translocation in cyanobacteria; MSX - methionine sulfoximine; AOA - amino oxyacetate; PMSF - phenylmethyl-sulfonyl fluoride

STEWART 1980b, LEA et al. 1982, MEEKS 1981, FLORES et al. 1983b). Glutamate dehydrogenase activity was detected at low levels in some cyanobacteria but not in most strains assayed (STEWART 1980b). There is almost general agreement that this enzyme plays a negligible role (if any) in primary ammonium assimilation in most cyanobacteria. Though glutamate dehydrogenase has been reported as the main ammonium assimilating pathway in *Anacystis nidulans* strain UTEX 625 when grown on nitrate (MEEKS et al. 1977, MEEKS 1981), GS-GOGAT appears to be the main route in nitrate-grown *Anacystis nidulans* strain 1402-1 (FLORES et al. 1983b). Attempts to detect glutamate dehydrogenase activity in some strains of *Anacystis* (HOARE 1967, FLORES et al. 1983b) have been unsuccessful. The involvement of alanine dehydrogenase in the primary assimilation of ammonium by cyanobacteria remains a matter of discussion (STEWART 1980b). Results from ^{13}N tracer studies (MEEKS et al. 1977, THOMAS et al. 1977, MEEKS 1981) point to its active participation in *Cylindrospermum licheniforme, Anabaena cylindrica, Plectonema boryanum,* and *A. nidulans* strains UTEX 625, but not in *Anabaena variabilis*. Carbamyl phosphate synthetase does not seem to significantly contribute to ammonium assimilation by cyanobacteria (MEEKS et al. 1977).

3 Uptake mechanisms

Free-living cyanobacteria inhabit environments that contain in the main low dissolved concentrations of the nutrients required in largest quantity for growth, such as carbon dioxide and nitrogen sources. Most filamentous forms capable of dinitrogen fixation carry out this process in specialized cells termed heterocysts, although a few nitrogen-fixing forms in which no such differentiation occurs are also known. It is rarely possible to detect extracellular ammonia in nitrate- or dinitrogen-using cultures, although the intracellular pool of ammonia in all cases where it has been measured is of the order of 1 mM (BOUSSIBA et al. 1984c, RAI et al. 1984, KERBY et al. 1986, RITCHIE and GIBSON 1987a, b). In symbiotic forms, in contrast, upwards of 60% of the total products of $^{13}N_2$ reduction formed in a five-minute incubation of freshly isolated *Azolla* symbionts appeared as $^{13}NH_4^+$ in the surrounding medium (MEEKS et al. 1985), indicating that the limiting membrane of these cyanobacteria must be permeated by at least one of the two possible forms of ammonia, NH_3 or NH_4^+. As discussed further below, the extensive production of NH_4^+ from both exogenous and endogenous sources by other cyanobacteria incubated under special conditions suggests that this is a general phenomenon. It has generally been assumed that specialized channels or carriers would be required to allow significant movements of the ionic form NH_4^+ across any membrane, but that NH_3, a small, gaseous molecule, should be quite freely permeable. However, in very few cases has this assumption been supported by measurements. The only determination of NH_3 permeability in a cyanobacterium appears to be for *Synechococcus* R-2 (RITCHIE and GIBSON 1987b); the values for P of approximately 6 mm. sec^{-1} are very similar to values obtained for the charophyte *Chara corallina* (RITCHIE 1987).

Internal pools of total ammonia in cyanobacteria, as well as in other prokaryotes, generally amount to 0.6 - 1.5 mM (BOUSSIBA et al. 1984c, KALB et al. 1987, CORDTS and GIBSON 1987). The cyanobacterial cell is, however, divided into distinct intrathylakoid and cytoplasmic compartments, and a pH gradient across the thylakoid membrane in excess of 2 pH units has been estimated (FALKNER et al. 1976). If, as estimated in that study, the intrathylakoid compartment is 10% of the internal volume of the cell, and if the permeability of the thylakoids to NH_3 is similar to that of the cytoplasmic membrane, a significant proportion of the total ammonia pool extracted by acid would presumably be sequestered in the thylakoids while cells are illuminated.

More detailed work on the ammonium/methylammonium transport system of *Escherichia coli* that has appeared since the comprehensive reviews of KLEINER (1981, 1985) has demonstrated that this *amt* system is under the control of the general nitrogen regulatory system *ntr* (JAYAKUMAR et al. 1986). The physiological effects of transposon insertions strongly suggest that a single system is responsible for ammonium translocation in *E. coli* (JAYAKUMAR et al. 1989); this conclusion was also drawn from investigations with a mutant of *Klebsiella pneumoniae* (CASTORPH and KLEINER 1984). However, reliance upon a single system for ammonium transport does not appear to be a general rule among microorganisms. In addition to the case in cyanobacteria discussed later in this review, more than one system for NH_4^+ translocation appears to be present in *Rhodobacter sphaeroides* (CORDTS and GIBSON 1987). Multiple systems, generally defined genetically, have been found in many eukaryotic microorganisms, including *Aspergillus nidulans* (COVE 1979), *Saccharomyces cerevisiae* (DUBOIS and GRENSON 1979), *Neurospora crassa* (DUNN-COLEMAN et al. 1984) and *Chlamydomonas reinhardtii* (FRANCO et al. 1988).

3.1 Uptake of NH_3/NH_4^+

One of the characteristic features of active transport is accumulation of the solute against its concentration gradient. When the solute is an ion, then the magnitude and sign of the membrane potential must also be considered. We have shown that the unicellular *Anacystis nidulans* (now renamed *Synechococcus*) can accumulate ammonia against a concentration gradient of close to 1000, leading to an internal pool of 1 - 1.5 mM, while the external concentration was reduced to below the detection limit of approximately 2 µM (BOUSSIBA et al. 1984c). The membrane potential maintained by the cells under comparable conditions was approximately -120 mV, implying significantly higher internal pools than could be accounted for by electrical equilibration (RITCHIE and GIBSON, unpublished). The pool was slightly reduced, but not lost, when cells were maintained in darkness for over 30 minutes, and the membrane potential was also maintained almost unchanged when cells were no longer illuminated. Ammonia was taken up at the same rate from solutions at pH 6.6 - 8.7 (Table 1), with no detectable change in affinity; an apparent K_m of less than 1 µM was inferred from these experiments. It is reasonable to conclude that the ammonia species transported must be NH_4^+, since its relative concentration is changed by less than 50% by proton dissociation over the pH range studied, while NH_3 would constitute only about $5x10^{-3}$ of the total ammonia present at pH 6.6. The pH insensitivity of ammonia uptake found in our experiments (BOUSSIBA et al. 1984c, ZIMMERMAN and BOUSSIBA 1986) contrasts sharply with the findings on what are thought to be ammonium /methylammonium systems studied with the analog in other microorganism (KLEINER 1985).

Table 1. Lack of effect of pH on ammonia uptake rate in *Anacystis nidulans* R-2

pH	Uptake rate (nmol min^{-1} mg^{-1} protein)
6.2	26.7
7.0	34.1
8.1	29.7
8.7	31.5

The effects of inhibitors on ammonia uptake suggest that amidation is critical in regulating net uptake. Compounds such as ionophores and dicyclohexylcarbodiimide (DCCD) affect energy conservation primarily, but indirectly will also reduce energy available for membrane transport as well as CO_2 fixation and the supply of acceptor for ammonia assimilation. Methionine sulfoximine, by inhibiting glutamine synthetase also prevents further net ammonia uptake. It is not possible in such experiments to distinguish between inhibition of the transport system itself and an indirect effect exerted through glutamine synthetase, while flux through the transporter continues unabated.

These considerations lead to the conclusion that the major form of ammonia crossing the cell membrane of neutrophilic cyanobacteria growing on this nitrogen source in the pH range 7-8 is primarily NH_4^+, and that some form of active transport must be involved in its accumulation. In alkalophilic strains such as *Spirulina platensis* and *Cyanospira rippkae*, which grow optimally at pH over 10 using ammonia as nitrogen source, a simple mechanism of NH_3 diffusion followed by trapping through protonation in the more acidic internal compartment may be sufficient. Recent data obtained using *S. platensis* (BOUSSIBA 1989) support this suggestion. At pH 10, uptake of ammonia consisted of two phases: an initial rapid disappearance of added ammonia, lasting less than 5 sec (the shortest time measured), followed by a slower phase (Fig. 2). The fast phase was unaffected when glutamine synthetase was completely inhibited by MSX. The slow phase, in contrast, was eliminated by MSX. Preliminary results (BOUSSIBA and BELKIN 1991) indicate that at an external pH of 10.0, the cytoplasmic pH is

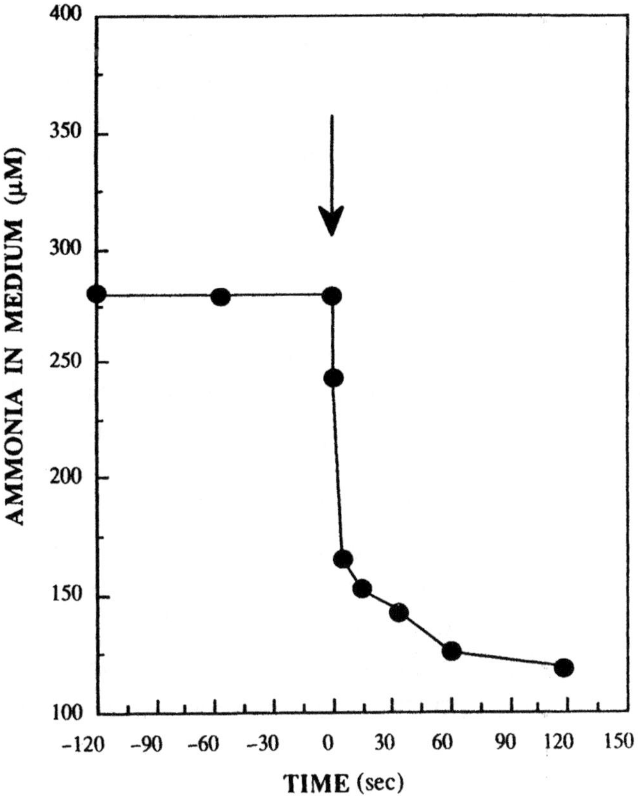

Fig. 2. Kinetics of the uptake of ammonia by *Spirulina platensis* at pH 10.0. Cells (to 1.8 mg protein ml^{-1}) were added at time zero (arrow). Data points indicate amounts of ammonia in the medium after removal of cells

approximately 8.5, as has been found also for other alkalophilic bacteria (KRULWICH and GUFFANTI 1989). The ΔpH is thus large enough and of the correct sign to lead to an intracellular accumulation of ammonia, following diffusion of NH_3 (GAENSSLEN and MCCARTY 1971, SCHULDINER et al. 1972). Indeed when cells were treated with toluene to eliminate any pH gradient, the fast influx of ammonia was abolished. Following influx, the continuous activity of glutamine synthetase supports further net entry by withdrawing ammonia from the internal pool.

3.2 Methylammonium uptake

As has been the case for studies of ammonium uptake, the major interest in the investigations that have made use of the analog methylammonium is in connection with broader aspects of nitrogen metabolism and its regulation. Much emphasis has therefore been placed on correlating methylammonium uptake with nitrogen fixation and with intracellular amidation carried out by glutamine synthetase.

3.2.1 Effect of growth conditions

Cyanobacteria grown either with nitrate or with dinitrogen as nitrogen source have regularly been found to take up $CH_3NH_3^+$ from suspending media buffered at approximately neutral pH. Observed uptake rates are up to one tenth of the rate at which NH_4^+ can be removed from the medium. However, growth with ammonia represses methylamine uptake to less than 10% of the rate measured in cells grown on other nitrogen sources, although such cells remain able to remove ammonia to below measurable concentrations. This response of $CH_3NH_3^+$ uptake to growth environment in itself provides a strong argument for existence of a specific transport system.

3.2.2 Assay conditions

The pH of the suspending medium governs the extent to which an amine is protonated, and therefore the proportion of the total molecules available for transport as the ion or through passive diffusion. Many investigators have therefore chosen to work at external pH values of 7.0–7.5, where only 0.1% of the total methylamine added is uncharged, or in the range of 9 - 10, where around 50% is in this form. In other respects, the assay conditions used have differed widely; for example, most investigations with *Anabaena* species from STEWART'S (e.g., RAI et al. 1984, KERBY et al. 1987, REGLINSKI et al. 1989) and SINGH'S (SINGH et al. 1985, SINGH et al. 1986) laboratories used cells from nitrogen-fixing cultures washed and suspended in buffer only, while growth conditions were used in work with *A. flos-aquae* (TURPIN et al. 1984), and buffers supplemented with bicarbonate have been used for *Synechococcus* (BOUSSIBA et al. 1984a, BOUSSIBA and GIBSON 1985). The differences in carbon supply to cells may influence uptake kinetics, as is further discussed below.

Both equilibration and uptake measurements have commonly been carried out under illumination, and although some uptake may occur in darkness, there is general agreement that significant uptake of methylammonium at neutral, as opposed to alkaline pH requires energy, and is driven by the membrane potential ($\Delta\Psi$) component of protonmotive force. In *Synechococcus*, for example, $CH_3NH_3^+$ uptake was as rapid at pH 7.0 as at 7.5, where external and intracellular pH are equal, while transient reduction of $\Delta\ \Psi$ by adding high concentrations of K^+, or more permanent collapse by nigericin, is strongly inhibitory (BOUSSIBA et al. 1984a).

In *Synechococcus*, $CH_3NH_3^+$ uptake requires the presence of Na^+ in the external medium, but the precise role played by this ion in transport has not been defined. In common with most other types of cells, Na^+ is extruded from energized cyanobacteria (PASCHINGER 1977), and the very rapid flux in illuminated cells establishes a sodium motive force (ΔpNa^+) that exceeds ΔpH^+ by 15 to 20 mV (RITCHIE and GIBSON unpublished). Coupled movements of $CH_3NH_3^+$ and reentering Na^+ in response to the membrane potential is a possible explanation. Published experiments with filamentous cyanobacteria have been carried out in sodium-containing media, so that a specific requirement for this ion if it is indeed common to all cyanobacteria would not have been observed. A further difference between the uptake of $CH_3NH_3^+$ and of NH_4^+ is that a requirement for Na^+ cannot be shown for the natural substrate.

3.2.3 Kinetics

With a small number of exceptions, noted below, uptake of methylammonium by both filamentous and unicellular cyanobacteria is biphasic, an initial rapid phase which lasts about one minute being followed by a slower, sustained uptake which may continue linearly for at least one hour (RAI et al. 1984, BOUSSIBA et al. 1984a, SINGH et al. 1985, TURPIN et al. 1984). The rapid phase shows saturation kinetics, with apparent K_m values usually around 10 µM (Fig. 2), while the slow phase appears little affected by concentration at neutral pH in *Synechococcus*, although it is increased at pH 9 or above (KERBY et al. 1987, TURPIN et al. 1984). When excess NH_4^+ or a higher concentration of $^{14}CH_3NH_3^+$ is added to the suspension at the end of the rapid phase, all the radioactivity is released from the cells. However, after the slow phase has continued for some time, only a quantity of radioactivity approximately equal to that taken up in the rapid phase is released. In *Synechococcus* (BOUSSIBA et al. 1984a) and in *A. variabilis* (RAI et al. 1984, BOUSSIBA et al. 1984b, KERBY et al. 1987) the proportion of radioactivity retained corresponds accurately with the accumulation of γ-methyl glutamine within the cells. Since this accumulation is prevented in cells that have been exposed to methionine sulfoximine, the slow phase in these organisms appears clearly associated with the glutamine synthetase activity, and the ability of this enzyme to use the analog in place of ammonia. Different conclusions have been reached as a result of experiments with *A. flos-aquae* (TURPIN et al. 1984) and *A. doliolum* and *A. cycadae* (SINGH et al. 1985, SINGH et al. 1986, RAI et al. 1986), in part on the basis of experiments with MSX and its effect on the slow phase of uptake. An interesting glutamine auxotroph of *A. cycadae* (SINGH et al. 1985) isolated on the basis of resistance to MSX had no detectable glutamine synthetase activity as measured by the transferase assay, yet showed a slow phase of uptake comparable to that seen in the parent. Only a portion of the total radioactivity associated with the cells which had accumulated $^{14}CH_3NH_3^+$ for an extended time was chased out of the cells by adding NH_4^+ to the suspension, so that the possibility that the analog had been metabolized by the mutant, perhaps through some pathway independent of glutamine synthetase, has not been eliminated. Much of the discussion revolves around the effect and specific action of MSX which may have other targets in addition to glutamine synthetase, although it has been shown to be highly specific for this enzyme, with which it forms a covalent adduct (RONZIO and MEISTER 1968). Both CANVIN'S and SINGH'S groups have suggested a direct effect on a second specific transport system responsible for the slow phase of uptake. The interpretation of these experiments is complicated by the fact that MSX added to intact cells has itself to be transported into the cell to reach glutamine synthetase, and appears to use a transport system which also functions with glutamine (CHAPMAN and MEEKS 1983). Hence the onset of enzyme inhibition is slow relative to $CH_3NH_3^+$ uptake (see BOUSSIBA et al. 1984a, TURPIN et al. 1984). Since ammonia evolution sets in with a comparable time course, it has also been suggested that the considerably delayed inhibition of $CH_3NH_3^+$ uptake by MSX in *Synechococcus*, which affects the rapid phase of uptake also, is a result of net internal NH_4^+ generation when glutamine synthetase is inactivated (BOUSSIBA et al. 1984a).

Two uptake phases are not always observed. The rate of the second phase of uptake varies in different strains or species, and KERBY et al. (1987) noted that it was much slower in *A. cylindrica* than in *A. variabilis*. Although free-living strains of *A. cycadae* clearly show both rapid and slow phases of uptake, cyanobionts harvested freshly from host roots took up $CH_3NH_3^+$ rapidly at first, but no slow phase was observed; all accumulated radioactivity was released by added NH_4^+ (RAI et al. 1986), so that, although glutamine synthetase was active in broken cells, and was able to react with $CH_3NH_3^+$ in place of NH_4^+, this reaction was not occurring in the cells. Again, some different interpretations of the experiments is possible: the cells used in both of these investigations were incubated in buffers without bicarbonate, so that the availability of glutamate as an acceptor in the glutamine synthetase reaction may have been rate-limiting. A second suggestion is that the relative affinities of glutamine synthetases for $CH_3NH_3^+$ and NH_4^+ may vary in different strains, or even be dependent on growth conditions; some preliminary support for the former suggestion has been obtained (KERBY et al. 1987), and the observation that the cyanobionts were also unable to take up NH_4^+ could also be the result of regulatory effects on glutamine synthetase within the cell, that were relieved after extraction. It has been shown repeatedly that cyanobacterial glutamine synthetase is not regulated by the adenylylation/deadenylylation system found in enteric bacteria; however, activation of the enzyme by light-reduced thioredoxin has been demonstrated (PAPEN and BOTHE 1984). Preliminary experiments with antiserum prepared against *Synechococcus* strain R-2 glutamine synthetase have shown at least two electrophoretically separable forms of the native enzyme, whose ratio is affected by the nitrogen source used for growth (BRAKE and GIBSON, unpublished), and it is evident that the regulation of the enzyme *in vivo* is not yet fully understood.

3.2.4 Regulation

It has already been noted that growth of all the strains of cyanobacteria used in these investigations with ammonia as nitrogen source yields cells that fail to take up $CH_3NH_3^+$ rapidly at neutral pH. Nitrate-grown, and therefore uptake-competent, *Synechococcus* cells lose this activity when incubated in growth medium either with NH_4^+ or $CH_3NH_3^+$ although in the latter the cells become chlorotic and the increase in cell mass is limited. In contrast, the rate of transport of $CH_3NH_3^+$ is maintained and even enhanced if the medium contains glutamine, which in this strain supports only slow growth, presumably because of transport limitation (BOUSSIBA and GIBSON 1987). The interpretation of derepression kinetics in unicellular forms, as is the case with diazotrophic strains, is complicated by the accompanying nitrogen deficiency and need for synthesis of enzymes for utilizing the new nitrogen source, whether this be nitrate or dinitrogen. However, a rise in $CH_3NH_3^+$ uptake becomes apparent within one to two hours, and is dependent on new protein synthesis (BOUSSIBA and GIBSON 1987, RAI et al. 1986). These experiments indicate a coordinated response between transport and the utilization of nitrogen sources other than ammonia, but details of the regulatory system remain to be established. However, two recent publications bear on the genetic control of $CH_3NH_3^+$ transport. Mutants of *A. variabilis* selected for resistance to concentrations of $CH_3NH_3^+$ toxic to the parent at pH 7, had greatly reduced fast and slow phases of uptake at pH 7, although diazotrophic growth was unimpaired (REGLINSKI et al. 1989). This mutant therefore appears to lack the transport system responsible for rapid uptake, but not to suffer any ill effects from this lesion in terms of its ability to reduce and incorporate dinitrogen. These observations appear inconsistent with the suggested role of the transport system in recapturing internally generated NH_4^+ (KLEINER 1985, BOUSSIBA et al. 1984a). It must be remembered, however, that the primary product of nitrogen fixation is assimilated into organic compounds within the heterocysts, rather than being transferred as such into the vegetative cells which are presumably the main site of $CH_3NH_3^+$ transport. Considerable differences in detailed function between differentiated diazotrophs and unicellular cyanobacteria using nitrate, and experimental definition of $CH_3NH_3^+$ transport in filamentous cyanobacteria grown with nitrate does not appear to have been attempted. An unexpected finding with the mutant was that growth in the presence of $CH_3NH_3^+$ appeared to result in expression of a further transport system for this compound although, in common with all cyanobacteria, $CH_3NH_3^+$ will not support growth of the mutant and was metabolized no further than to γ-methyl glutamine. A mutant of *Nostoc muscorum*, resistant to the herbicide glyphosate, had lost the ability to differentiate heterocysts, to fix dinitrogen, and also to accumulate $CH_3NH_3^+$ (SINGH et al. 1989). Some general regulatory phenomenon appears therefore to be involved in coordinating responses to changes in nitrogen supply, perhaps analogous to the *ntr* system of enteric bacteria (JAYAKUMAR et al. 1986).

4. Regulation of key enzymes in nitrogen assimilation

4.1 Ammonium repression of nitrate assimilation

When supplied with both ammonia and nitrate as potential nitrogen sources in the growth medium, cyanobacteria will first completely deplete the medium of ammonia and only then utilize nitrate (OHMORI et al. 1977). It may be inferred that ammonia regulates key enzymes involved in nitrate assimilation. A direct effect on the activity or synthesis of nitrate reductase has been ruled out, since reduction of nitrate uptake by whole cells occurred almost immediately on adding ammonia, and the activity of nitrate reductase measured in permeabilized cells that had been exposed to ammonia for short periods was barely affected. An alternative explanation for inhibition of nitrate uptake through a direct effect of ammonia on nitrate transport has been suggested (OHMORI et al. 1977, RAI et al. 1981, FLORES et al. 1980, 1983). The inhibition of nitrate uptake by ammonia is eliminated if glutamine synthetase, which in cyanobacteria appears to constitute the main, if not the only, route of assimilation into organic linkage (OHMORI and OHMORI 1988, GUERRERO and LARA 1988), is blocked (RAI and SINGH 1982) by the specific inhibitor methionine sulfoximine (MSX) (RONZIO and MEISTER 1968, STEWART and ROWELL 1975). Thus ammonia inhibition of nitrate uptake requires continued assimilation of internal ammonia by glutamine synthetase. This situation is comparable to the effect of MSX on nitrogenase in the presence of ammonia: the addition of ammonia to an uninhibited nitrogen-fixing culture represses nitrogenase (STEWART and ROWELL 1975), but nitrogenase activity is maintained if MSX is also added. Based on such findings, it has been suggested that glutamine, or a closely related metabolite derived from that compound, mediates inhibition of both nitrate and dinitrogen utilization.

4.2 Nitrogenase regulation by ammonia

In free-living N_2-fixing bacteria, synthesis of nitrogenase is repressed after addition of ammonia. In most bacteria, but not some photosynthetic bacteria, nitrogenase in cell extracts remains active and following several generations, nitrogenase activity is lost due to degradation and dilution (DROZD et al. 1972, DAESCH and MORTENSON 1972, KURZ et al. 1975, TUBB and POSTGATE 1973).

The studies by FOGG (1949), demonstrating that ammonia represses the differentiation of heterocysts (the sites of nitrogenase in certain filamentous cyanobacteria) and inhibits N_2 fixation by *Anabaena* species have been confirmed by numerous studies on a variety of cyanobacteria (see for reviews STEWART 1980, CARR and WHITTON 1982). However, the mechanisms of regulation of nitrogenase biosynthesis and activity in heterocystous cyanobacteria are still largely unknown. Since ammonia or its metabolites repress the formation of heterocysts, it has been difficult to establish whether the inhibition is due to an inhibition of heterocyst differentiation or to a repression of nitrogenase biosynthesis.

Furthermore, ammonia may indirectly inhibit nitrogenase activity *in vivo* by interrupting substrate supply (OHMORI and HATTORI 1974, 1978, MURRY and BENEMANN 1979). In natural waters the concentrations of ammonia are well below those typically used in laboratory studies and it is unclear to what extent these findings directly relate to nitrogenase regulation in the environment (see HORNE and COMMINS 1987).

In addition to this long-term effect of ammonia on nitrogenase synthesis, a rapid and reversible inhibition of nitrogenase by ammonia has been termed "ammonia switch-off" and has been studied in a variety of photosynthetic bacteria. A covalent modification of the Fe protein was found to inactivate nitrogenase following addition of NH_4^+ (JOUANNEAU et al. 1983, KANEMOTO and LUDDEN 1984, HARTMANN et al. 1986). This "switch-off" is abolished in certain bacteria treated with MSX implicating glutamine as a metabolite of regulatory importance in this phenomenon (ZUMFT and CASTILLO 1978, ARP and ZUMFT 1983, CEJUDO et al. 1984, HARTMANN et al. 1986). Methylamine mimics the "ammonia switch-off" in *Rhodopseudomonas capsulata* and γ-methyl glutamine, like glutamine, was thought to be responsible for nitrogenase inactivation (YOCH et al. 1983). A rapid inactivation of nitrogenase in the cyanobacterium *A. variabilis* by ammonia has been reported to occur at pH 10 but not at pH 7.2 (REICH et al. 1986). However, MSX only partially protected nitrogenase activity at pH 10. In cyanobacteria evidence is required that the rapid "ammonia switch-off" is due to a covalent modification of the Fe protein following ammonia addition, as has been found in other N_2-fixing bacteria.

5 Biotechnological aspects

5.1 Ammonia as the preferred nitrogen source for growth

Cyanobacteria can utilize a number of nitrogen sources to support growth. All can use nitrate or ammonia when available, and some are also able to fix atmospheric nitrogen. When fixed nitrogen is available, this is almost always used in preference to dinitrogen.

Nitrate or dinitrogen must first be reduced to ammonia before assimilation, and the preferential use of ammonia could thus be accounted for in terms of cell energy balance. The high energetic cost of dinitrogen reduction has been extensively discussed, and reduction of nitrate must also involve a significant diversion of reductant from carbon dioxide fixation. We have recently determined that *Anabaena siamensis*, a nitrogen-fixing cyanobacterium isolated from rice paddies in Thailand grew faster under limiting illumination when using ammonia than with nitrate or under nitrogen-fixing conditions, with doubling times of 3.7 h and 5.6 h, respectively (BOUSSIBA and THOMAS, unpublished).

Today most enterprises cultivating microorganisms in large scale utilize ammonia as the preferable N-source since it is the cheapest N-source. Furthermore, this cation was reported to preclude the proliferation of zooplankton in outdoor algal cultures. Indeed, we have looked at this phenomenon, the addition of 5 to 10 mM NH_4^+ to cultures affected by zooplankton causes a drastic reduction of these microorganisms (BOUSSIBA et al. 1987, BOUSSIBA et al. 1988). Therefore, we suggest to use ammonia as the ultimate N-source not only because of its lowest price as N-source but also because it may prevent contamination of zooplanktonic microorganisms.

5.2 Toxic effects of ammonia at high pH

The preponderance of NH_3 over NH_4^+ at pH values greater than about 9.2, the pK_a for ammonia at 20-30°C, and the relatively high permeability of the unionized form of ammonia, posed a potential hazard for photosynthetic microorganisms in which pH gradients play a key role in energy conservation. Inhibition of photophosphorylation, associated with elevated intrathylakoid pH and decreased transmembrane pH gradients, has been thoroughly documented in isolated chloroplast thylakoids.

Ammonia has also been reported to uncouple cyanobacterial photosynthesis at high pH (ABELIOVICH and AZOV 1976), since 5 mM ammonia severely inhibited $^{14}CO_2$ photoassimilation at pH 9 in a number of strains. The degree of inhibition appears, however, to be much reduced in strains that are restricted to growth in alkaline environments, and so may be considered alkalophilic. Fig. 3 shows that while 5 mM ammonia inhibited light-dependent O_2 evolution by 90% in a neutrophilic *Anabaena* species, only 30% inhibition was observed in the alkalophilic *Spirulina platensis* at the same pH. These observations may be due to the different pH gradients maintained by the two organisms; the much greater gradient in *Anabaena*, which maintains an average internal pH of about 7.5, as compared with 8.5 in *Spirulina* (BELKIN and BOUSSIBA 1991) might be expected to result in a larger influx of NH_3 in the neutrophile and hence greater disruption of energy conservation.

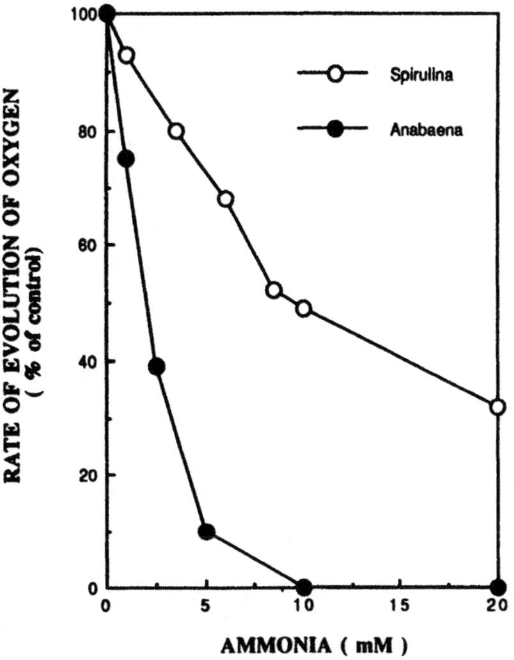

Fig. 3. Effects of ammonia on the light-dependent evolution of oxygen at pH 10.0. Both *S. platensis* and *Anabaena* sp. were present at 2.5 mg Chl ml^{-1}. Control rates of evolution of oxygen (without added ammonia) were 6.36 and 4.52 mol O_2 mg^{-1} Chl min^{-1} for *Spirulina* and *Anabaena*, respectively

This phenomenon is used to explain one of the most peculiar events in which *Spirulina* outdoor cultures are contaminated with the green alga *Chlorella*. The way to deal with this situation is to add to the growing *Spirulina* cultures maintained usually at high pH's (9.5-10.0) 2-3 mM ammonia. This addition is causing immediately the deterioration of the *Chlorella* cells in the *Spirulina* cultures. A possible explanation to this event is the toxicity of ammonia to the *Chlorella* cells caused by the larger influx of this molecule due to the lower internal pH (as in the case of the *Anabaena* sp., see above). As a consequence of this influx the internal thylakoid pH is elevated and collapse the ΔpH existing between the two sides of the thylakoids required for ATP synthesis, leading to the disintegration of the *Chlorella* cells. We suggest therefore to use ammonia as N-source in *Spirulina* cultures to prevent such events and also to reduce the proliferation of other microorganisms.

5.3 Large scale production of nitrogen-fixing cyanobacteria for rice farming

The use of nitrogen-fixing cyanobacteria is currently confined mainly to their utilization as nitrogen fertilizers for rice production in tropical countries. It is generally assumed that the nutrients fixed by cyanobacteria are made available mainly in form of ammonia to the rice plants through exudation autolysis and microbial decomposition. (ROGER and KULASOORIYA 1980, VENKATARAMAN 1986, ROGER et al. 1992). However, this biofertilization technique is still limited mainly due to the inability to efficiently produce good quality inocula and the difficulty to reestablish the inoculated strains in the field (ROGER and WATANABE 1986, GRANT et al. 1986). Recently some progress has been achieved in the production of some rice field isolates in open raceway reactors (FONTES et al. 1987, BOUSSIBA 1988, QUERIJERO-PALACPAC et al. 1990, BOUSSIBA 1991, QIAN-LIN et al. 1991). In most of these cases the cultures had to be kept of relatively low cell density to obtain high yield. Application of the biomass harvested from the pond to the field required processing, i.e., harvesting and drying, which reduced cell viability and increased the cost of production (WATANABE 1984, ROGER et al. 1987).

One way to overcome some of the above mentioned problems is to try to cultivate the cyanobacteria in a reactor that ensures high yield when grown at a relatively high cell density. In such a case, direct application of the cyanobacteria to the field may be considered. Recently a new kind of reactor - a closed tubular reactor-is being developed in our laboratory to overcome some of the problems of low productivity caused mainly by suboptimum temperatures in open raceway ponds (VONSHAK and RICHMOND 1988). The efficiency of this bioreactor for the production of a nitrogen-fixing cyanobacterium rice field isolate, *A. siamensis*, was tested. We have shown (BOUSSIBA 1993) that the rate of biomass production is significantly higher in this type of reactor, as compared to the open pond, i.e., 0.31 g l^{-1} and 0.09 g l^{-1}, respectively. This system may thus become a reliable means for supplying desired microorganisms, which are in some cases major components in a food-chain production system (BOUSSIBA et al. 1988). Moreover, substituting nitrogen biofertilizers with nitrogen-fixing cyanobacteria produced in a closed tubular bioreactor has other specific advantages. Direct application of the dense cultures into the field may reduce the cost of production, since no harvesting or drying process will be required. It might also improve cell viability, which drastically drops when these algae are dried before application to the field. Using a closed tubular bioreactor will also prevent dilution by rainfall of the growing cultures (or even total cell wash out), which often occurs in shallow open ponds. Finally, the prolonged period of production possible with closed bioreactor can provide a continuous supply of a good source of protein to be used for enterprises other than rice cultivation, such as fish or chicken production, and may help reduce the migration of farmers to urban areas when the rice season is over. The production of nitrogen-fixing rice field cyanobacteria in this kind of tubular reactors for rice farming has thus great potential and may considerably enhance this fertilization technique used in tropical countries where rice is considered a major staple for the human diet.

5.4 Ammonium excretion

Cyanobacteria can take up ammonia from their environments at a rate that appears to be limited less by the activity of their transport systems for NH_4^+ at neutral pH or by diffusion of NH_3 at alkaline pH than it is by the net rate at which the internal pool of ammonia is converted into amino groups. The internal pools appear relatively constant (e.g., BOUSSIBA et al. 1984c, BOUSSIBA et al. 1984b), suggesting that factors leading to internal ammonia production will also result in net release of this compound. Substantial production of ammonia by cyanobacteria can indeed be observed under special conditions, and the biotechnological implications for tropical agriculture have received considerable attention.

5.4.1 The mechanism of ammonium excretion

A common feature of these experimental systems is that glutamine synthetase is impaired, while energy supply and alternate nitrogen sources continue to be available. One may ask whether ammonia is lost from the cells as NH_4^+, possibly through reversal of the transport system, or through diffusion of NH_3, which, as discussed above, is able to diffuse quite rapidly through the cell membrane. Indirect evidence obtained recently suggests that ammonia is lost from the cell as NH_3 in a process that is not energy-dependent. In *Spirulina platensis* the loss of ammonia was pH dependent, the rate increasing when the external pH was more acid. Ammonia release brought on by MSX inhibition was unaffected by pretreatment of the cells with N-ethylmaleimide, which inhibited ammonia uptake without affecting energy balance over a period of at least one hour; this may indicate that uptake and release are independent processes, and that excretion is not mediated by channels involved in NH_4^+ uptake. In contrast to the mutants of enteric bacteria that lack the *amt* system (JAYAKUMAR et al. 1989, CASTORPH and KLEINER 1984) and release ammonia, a recently isolated mutant from a nitrogen-fixing cyanobacterium (BOUSSIBA and THOMAS, unpublished) released ammonia continuously while fixing nitrogen, but still possesses an ammonium uptake system as evidenced by ability to accumulate $^{14}CH_3NH_3^+$ Taken together, the weight of experimentation suggests that ammonia leaks from cells because of the inherent permeability of NH_3, and that no specific carrier is involved in its release.

5.4.2 Conditions favoring ammonia excretion

Ammonia formation is generally observed when assimilation is blocked by pretreatment of cells with MSX to covalently inhibit glutamine synthetase. Nitrate supplied to such cells is reduced to ammonia and released (GUERRERO et al. 1982, RAMOS et al. 1982a, 1982b), and cells remain active for long periods. Substantial ammonia production from dinitrogen has also been observed in the presence of MSX (MUSGRAVE et al. 1982, RAMOS et al. 1983, RAMOS et al. 1987, HALL et al. 1985, BROUERS and HALL 1985, NEWTON and CAVINS 1985. Lower rates of MSX-induced ammonia production resulting from protein breakdown in strains incapable of nitrogen fixation have also been reported (BOUSSIBA et al. 1984b, BERGMAN 1984). A number of mutants of nitrogen-fixing strains that release ammonia continuously have been isolated. The most common features of these mutants are reduced glutamine synthetase activity together with increased nitrogenase (BOUSSIBA et al. 1984b, POLUKHINA et al. 1982, LATTORE et al. 1986, SPILLER et al. 1986, SUBRAHMANIAN and SHANMUGASUNDARAM 1986).

5.4.3 Ammonia production by immobilized cyanobacteria

Light driven synthesis of ammonia from dinitrogen and water via photosynthesis is an interesting biotechnological application for conversion of solar radiation into chemical energy from readily available substrates; moreover, the resulting compound, ammonia, can be used as a fertilizer.

Heterocystous cyanobacteria are ideal organisms for such studies; they are O_2 evolving photosynthesizers able to fix atmospheric nitrogen via ATP dependent nitrogenase activity (CARR and WHITTON 1982). Although nitrogenase activity is inhibited by O_2, heterocystous cyanobacteria are able to fix N_2 in aerobic conditions because their nitrogenase is located in the heterocyst which provides O_2 protection for the enzyme (heterocysts lack photosynthetic O_2 evolution and have a cell wall envelope which serves as an O_2 barrier). Under normal physiological conditions, the nitrogen fixed in the heterocyst is transported to the adjacent vegetative cells as glutamine which then enters into the cellular metabolic pool. The conversion of NH_4^+ to glutamine is catalyzed by the enzyme glutamine synthetase located in the heterocyst.

The main strategy to induce extracellular release of ammonia is to supply a specific inhibitor of glutamine synthetase activity, e.g., the glutamate analog L-methionine-DL-sulfoximine (MSX) (STEWART and ROWELL, 1975). The continuous presence of the inhibitor was a limitation for long term ammonia production and thus for practical applications; this is because MSX leads rapidly to a deficiency of glutamine and other nitrogenous compounds which are necessary for metabolism of the cell with cessation of ammonia production and finally lysis of the cells. It was shown possible to overcome at least partly this limitation and to lengthen production period by adding small amounts of glutamine (RAMOS et al. 1984) or by allowing the cells to recover in the absence of inhibitor (RAMOS et al. 1984, BROUERS, 1986). Another approach for increasing the long term productivity is the use of immobilized resting cells (MUSGRAVE et al. 1982, HALL et al. 1985).

A number of studies have been done in recent years on the immobilization of photosynthetic organisms for the production of pharmaceuticals, chemicals, and fuels. Immobilization appears to offer several advantages such as increased biocatalytic capacity due to increased cell densities, stabilization of enzyme activities, the possibility of avoiding out of cells at high dilution rates, the possibility for continuous operation, and lower costs of isolation of products. Entrapment in porous cells and foams has become the most popular techniques for the immobilization of whole cell as they generally do not use toxic reagents. The most common used matrices for photosynthetic cells are agar and alginate gels and polyurethane and polyvinyl foams. A disadvantage of the use of agar and alginate is their low mechanical stability for use in bioreactors over the long term; moreover, Ca alginate gels cannot be stored or used in buffers able to leach out the Ca ions. This requirement precludes phosphate and citrate buffers. Polyurethane and polyvinyl foams offer better mechanical properties and are neutral to most commonly used ions.

5.4.4 Mutants which continuously release ammonia

Nitrogen-fixing cyanobacteria are being used as nitrogen biofertilizers in rice fields in countries where rice is the major stable diet (VENKATARAMAN 1975). Although some strains which thrive in rice fields release small quantities of the major fertilizing product - ammonia - during active growth, most of the fixed products are made available mainly through autolysis and microbial decomposition (MARTINEZ 1984). Under these circumstances, it is difficult to control the flow of nitrogen compounds needed for the development of rice plants. A possible solution to this problem is to develop strains of cyanobacteria which release ammonium continuously.

Mutants of *Anabaena variabilis* resistant to the ammonium analog, ethylene diamine (EDA), and to the L-glutamate analog, L-methionine-DL-sulfoximine (MSX), have been reported to release ammonium (KERBY et al. 1986, KERBY et al. 1987, POLUKHINA et al. 1982, SPILLER et al. 1986). They exhibit, however, a slower growth rate as compared with that of their parents. Another major problem of using cyanobacteria as biofertilizer is the competition between indigenous and introduced strains, the former generally dominating. It was assumed that ammonium-excreting mutants that had been isolated from strains indigenous to rice fields would overcome the constraints of the rice field environment better than strains derived from other habitats. The reinoculation and establishment of these mutants in rice fields would thus be comparatively more successful.

In this respect, *Anabaena siamensis* a rice field isolate from Thailand proved promising, and is already marketed as algal biofertilizer for rice fields. Its efficiency in increasing the growth and yield of rice plants is apparently due to its high nitrogen-fixing capacity (ANTARIKANONDA 1982a, 1982b). It was reported to release a variety of amino acids during active growth (ANTARIKANONDA 1984) but not ammonium as observed for other nitrogen fixers found in rice fields. Selection of mutants resistant to MSX was found to lead to the reduction in glutamine synthetase activity and consequently the unassimilated ammonium is released into the medium without the induction of MSX (SPILLER et al. 1986). It is assumed that a strain releasing ammonium continuously would be a better biofertilizer (LATTORE et al. 1986). The MSX-resistant mutant of *A. siamensis*, SS1 isolated by us, seems to conform with this expectation (THOMAS et al. 1990).

This mutant releases ammonium due to the high activity of nitrogenase, both being controlled by cell density of the culture. The direct effect is apparently light availability to each cell, which becomes progressively limited as cell density increases. The rate of ammonium release is consequently maximum only during the early-log phase of growth in batch cultures. A similar pattern is also seen under immobilized conditions. Based on the above observations, SS1 growing in continuous culture at low cell density (chlorophyll value of 5 to 7 $\mu g\ ml^{-1}$) seems an ideal system for sustained ammonium release. The rate of ammonium release by SS1 is lower than the rates obtained for other mutants of *A. variabilis*, i.e., 35 to 50 $\mu mol\ mg\ Chl^{-1} h^{-1}$ in batch and immobilized cultures (KERBY et al. 1986, SPILLER et al. 1986). No data are provided on the excretion of ammonium by these mutants under steady state growth conditions. Hence, comparison between rates of ammonium release by the various mutants is inadequate.

Although both the parent and SS1 strains have a similar pattern of nitrogenase activity, the activity of SS1 enzyme is 30% higher than that of the parent during steady-state growth and 50% higher under immobilizing conditions. In the presence of ammonium SS1 nitrogenase activity is about five-fold higher than that of the parent, due to a stronger repression by the end product in the parent. Glutamine synthetase (GS) also seems defective in SS1, exhibiting less than 50% of the parent's enzyme activity. Hence it appears that the decreased susceptibility of nitrogenase activity, to ammonia repression and the defective glutamine synthetase are the main factors controlling the ammonium excretion in SS1 mutant. The other ammonium-excreting mutants of *A. variabilis*, i.e., SA1, ED81 and ED92, were found to have derepressed nitrogenase activity and lower GS activity. Analysis of GS and its mRNA into EDA-resistant mutants (HIEN et al. 1988) suggested that ED92 is a regulatory mutant containing less GS mRNA and consequently less GS protein, as found for *A. azollae* growing in symbiosis (NIERZWICKI-BAUER and HASELKORN 1986). On the other hand, ED81 is a structural mutant with a catalytically deficient GS, resulting in reduced activity of GS, which synthesizes protein at equal amount to its parent (KERBY et al. 1987), as found for *Nostoc* sp. 7801 growing in symbiosis (JOSEPH and

MEEKS 1987, LEE et al. 1988). Immunoblot analysis of crude proteins and Western blot analysis of GS from SS1 reveals that its quantity and mobility (*Mr*=50,000) is the same as those in the parent. A modification similar to that of ED81 may also have occurred in SS1, forming a catalytically deficient GS.

Fertilization of rice plants under laboratory conditions by application of another MSX-resistant mutant of *A. variabilis*, SA1, was found successful (LATTORE et al. 1986). The shorter doubling time of SS1 and the lack of a lag period at the beginning of the growth cycle as compared with SA1 (SPILLER et al. 1986), which exhibits a lag period, are obvious assets for mass cultivation of the former. These characteristics make possible the production of considerable amounts of inoculum material within a short period. The usefulness of SS1 as a biofertilizer to rice plants in the actual isolate location, Thailand, should be studied in order to scale up the technology transfer from laboratory to the rice field.

6 Conclusions

Ammonia seems to be one of the most interesting inorganic molecule involved in many processes in growth physiology of cyanobacteria. The fact that this molecule, except for being the most preferable N-source, can regulate so many processes in N-metabolism, photosynthesis and is also involved in biotechnological aspects, has promoted vast number of studies related to this different topics. Todate it is well accepted that the route through which this molecule is assimilated is via the GS/GOGAT pathway. As for the mechanisms by which this molecule (NH_3/NH_4^+) enters or exits cyanobacterial cells, substantial evidence has been suggested to support the existence of a transport systems which can function in the accumulation and retention of ammonia, and of its analog methylamine, from neutral environments. The effective competition between ammonium and methylammonium ions in uptake experiments carried out at pH values around neutrality leads to the conclusion that the ions share a common uptake pathway. However, cells in which methylammonium transport is repressed are unimpaired in their ability to take up ammonium, and differences between ionic requirements for uptake of the two ions have also been found. The suggestion that multiple systems exist for transporting NH_4^+, and that only one of these also transports $CH_3NH_3^+$, has therefore been put forward, but requires substantiation for example by genetic evidence.

Clearly, molecular approaches will be invaluable in answering the questions outstanding, and in establishing the physiological significance of the transport systems as well as in attempts to exploit the potential of cyanobacteria for increasing soil fertility.

As for the biotechnological aspects related to this molecule such as production of ammonia in immobilized systems, it is obvious that economically there is no advantage in such production systems. Ammonia is a very cheap product and immobilized systems may be used to produce other cell compounds which have high commercial value such as amino acids or for other rare molecules of special interest.

However, the production of nitrogen-fixing cyanobacteria for N-fertilization in rice fields in relatively simple tubular reactors suggested by us may be a much more attractive way to utilize those nitrogen-fixing cyanobacteria. The selection of mutants which continuously release ammonia may also help to overcome some of the difficulties involved in this biotechnology.

References

Abeliovich A, Azov Y (1976) Toxicity of ammonia to algae in sewage oxidation ponds. Appl Environ Microbiol 31:801-806

Antarikanonda P (1982a) Effect of salinity on growth, nitrogen fixation and sodium uptake of rapidly growing N_2-fixing blue-green alga *Anabaena* sp. TA1. Microbios 34:177-184

Antarikanonda P (1982b) Influence of pretreating the seeds with extract of *Anabaena siamensis* on germination and growth of some rice varieties from Thailand. FAO Internl Commission 31:37-39

Antarikanonda P (1984) Production of extracellular free amino acids by cyanobacterium *Anabaena siamensis*. Curr Microbiol 11:191-196

Arp DJ, Zumft WG (1983) L-methionine-DL-sulfoximine as a probe for the role of glutamine synthetase in nitrogenase switch-off by ammonia and glutamine in *Rhodopseudomonas palustris*. Arch Microbiol 134:17-22

Belkin S, Boussiba S (1991) Resistance of *Spirulina platensis* to ammonia at high pH values. Plant Cell Physiol 32:952-958

Bergman B (1984) Photorespiratory ammonium release by the cyanobacterium *Anabaena cylindrica* in the presence of methionine sulfoximine. Arch Microbiol 137:21-25

Boussiba S (1988) N_2-fixing cyanobacteria as nitrogen biofertilizer. A study with the isolate *Anabaena azollae*. Symbiosis 6:129-138

Boussiba S (1989) Ammonium uptake in the alkalophilic cyanobacterium *Spirulina platensis*. Plant Cell Physiol 32:303-314

Boussiba S (1991) Nitrogen-fixing cyanobacteria - potential uses. Plant Soil 137:177-180

Boussiba S (1993) Production of the nitrogen-fixing cyanobacterium *Anabaena siamensis* in a closed tubular reactor for rice farming. Microb Releases 2:35-39

Boussiba S, Gibson J (1985) The role of glutamine synthetase activity in ammonium and methylammonium transport in *Anacystis nidulans* R2. FEBS Lett. 180:13-16

Boussiba S, Gibson J (1987) Regulation of methylammonium- ammonium transport in the unicellular cyanobacterium *Synechococcus* R-2, PCC 7942. FEMS Microbiol Lett 43:289-294

Boussiba S, Gibson J (1991) Ammonia translocation in cyanobacteria. FEMS Microbiol Rev 88:1-14

Boussiba S, Dilling W, Gibson J (1984) Methylammonium transport in *Anacystis nidulans* R-2. J Bacteriol 160:204-210

Boussiba S, Liu X-Q, Gibson J (1984) Endogenous ammonia production by *Anacystis nidulans* R-2 induced by methionine sulfoximine. Arch Microbiol 138:217-219

Boussiba S, Resch CM, Gibson J (1984) Ammonium uptake and retention in some cyanobacteria. Arch Microbiol 138:287-293

Boussiba S, Sandbank E, Shelef G, Cohen Z, Vonshak A, Ben-Amotz A, Arad S, Richmond A (1988) Outdoor cultivation of the marine microalga *Isochrysis galbana* in open reactors. Aquaculture 72:247-253

Boussiba S, Vonshak A, Cohen Z, Avissar Y, Richmond A (1987) Lipid and biomass production by the halotolerant microalgae *Nannochloropsis salina*. Biomass 2:37-48

Brock TD (1973) Evolutionary and ecological aspects of the cyanophytes. In: Carr NG, Whitton BA (eds) The Biology of blue- green algae, Blackwell Scientific Publications, Oxford, 487-500

Brouers M (1986) Hydrogen production by immobilized *Scenedesmus* cells and ammonia production by immobilized *Anabaena* filaments. In: Atkinson B, Black GM, Webb C (eds) Process engineering aspects of immobilized cell systems, The Inst of chemical engineers, Warwicks

Brouers M, Hall DO (1985) Ammonia and hydrogen production by immobilized cyanobacteria. J Biotechnol 3:307-321

Carr NG, Whitton BA (eds) (1982) The biology of cyanobacteria. Blackwell Scientific Publications, Oxford

Castorph H, Kleiner D (1984) Some properties of a *Klebsiella pneumoniae* ammonium transport negative mutant. Arch Microbiol 139:245-247

Cejudo FJ, de la Torre A, Paneque A (1984) Short-term ammonium inhibition of nitrogen fixation in *Azotobacter*. Biochem Biophys Res Commun 123:431-437

Chapman JS, Meeks JC (1983) Glutamine and glutamate transport by *Anabaena variabilis*. J Bacteriol 156:122-129

Ciferri O, Tiboni O (1985) The biochemistry and industrial potential of *Spirulina*. Annu Rev Microbiol 39:503-526

Codd GA, Stewart WDP (1973) Pathways of glycolate metabolism in the blue-green alga *Anabaena cylindrica*. Arch Microbiol 119:1-5

Cordts ML, Gibson J (1987) Ammonium and methylammonium transport in *Rhodobacter sphaeroides*. J Bacteriol 169:1632- 1638

Cove DJ (1979) Genetic studies of nitrate assimilation in *Aspergillus nidulans*. Biol Rev 54:291-327

Daesch G, Mortenson LE (1972) Effect of ammonia on the synthesis and function of the N_2-fixing enzyme system in *Clostridium pasteurianum*. J Bacteriol 110:103-109

Drozd JW, Tubb RS, Postgate JR (1972) A chemostat study of the effect of fixed nitrogen sources on nitrogen fixation, membranes and free amino acids in *Azotobacter chroococcum*. J Gen Microbiol 73:221-232

Dubois E, Grenson M (1979) Methylamine/ammonia uptake systems in *Saccharomyces cerevisiae*: multiplicity and regulation. Mol Gen Genet 175:67-76

Dunn-Coleman NS, Nassiff MD, Garrett RH (1984) Isolation and characterization of a methylammonium resistant mutant of *Neurospora crassa*. Curr Genet 8:423-427

Falkner G, Horner F, Werdan K, Heldt HW (1976) pH changes in the cytoplasm of the blue-green alga *Anacystis nidulans* caused by light-dependent proton flux into the thylakoid space. Plant Physiol 58:717-718

Flores E, Guerrero MG, Losada M (1980) Short-term ammonium inhibition of nitrate utilization by *Anacystis nidulans* and other cyanobacteria. Arch Microbiol 128:137-144

Flores E, Ramos JL, Herrero A, Guerrero MG (1983a) Nitrate assimilation by cyanobacteria. In: Papageorgiou GC, Packer L (eds) Photosynthetic prokaryotes: cell differentiation and function, Elsevier/North-Holland, Amsterdam, pp 363-387

Flores E, Romero J, Guerrero MG, Losada M (1983b) Regulatory interaction of photosynthetic nitrate utilization and carbon dioxide fixation in the cyanobacterium *Anacystis nidulans*. Biochim Biophys Acta 725:529-532

Fogg GE (1949) Growth and heterocyst production in *Anabaena cylindrica* Lemm. II. In relation to carbon and nitrogen metabolism. Ann Bot 13:241-259

Fontes AG, Vargas MA, Moreno J, Guerrero MG, Losada M (1987) Factors affecting the production of biomass by a nitrogen-fixing blue-green alga in outdoor culture. Biomass 13:33-43

Franco AR, Cardenas J, Fernandez E (1988) Two different carriers transport both ammonium and methylammonium in *Chlamydomonas reinhardtii*. J Biol Chem 263:14039-14043

Gaensslen RE, McCarty RE (1971) Amine uptake in chloroplasts. Arch Biochem Biophys 147:55-65

Gibson J (1981) Movement of acetate across the cytoplasmic membrane of the unicellular cyanobacteria *Synechococcus* and *Aphanocapsa*. Arch Microbiol 137:21-25

Grant IF, Roger PA, Watanabe I (1986) Ecosystem manipulation for increasing biological N_2 fixation by blue-green algae (Cyanobacteria) in lowland rice fields. J Biol Agric Hortic 3:299-315

Guerrero MG, Lara C (1988) Assimilation of inorganic nitrogen. In: Fay P, van Baalen C (eds) The cyanobacteria, Elsevier, Amsterdam, pp 163-176

Guerrero MG, Ramos JL, Losada M (1982) Photosynthetic production of ammonia. Experientia 38:53-58

Hall DO, Affolter DA, Brouers M, Shi, D-J, Yang L-W, Rao KK (1985) Photobiological production of fuels and chemicals by immobilized algae. Ann Proc Phytochem Soc Eur 26:161-185

Hartmann A, Fu H, Burris RH (1986) Regulation of nitrogenase activity by ammonium chloride in *Azospirillum* spp. J Bacteriol 165:864-870

Haystead A, Dharmawardene MWN, Stewart WDP (1973) Ammonia assimilation in a nitrogen-fixing blue-green alga. Plant Sci Lett 1:439-445

Hien NT, Kerby NW, Machray GC, Rowell P, Stewart WDP (1988) Expression of glutamine synthetase in mutant strains of the cyanobacterium *Anabaena variabilis* which liberate ammonia. FEMS Microbiol Lett 56:337-342

Hoare DS, Hoare SL, Moore RB (1967) The photoassimilation of organic compounds by autotrophic blue-green algae. J Gen Microbiol 73:317-326

Horne AJ, Commins ML (1987) Macronutrient controls on nitrogen fixation in planktonic cyanobacterial populations. New Zealand J Mar Freshwater Res 21:413-424

Jayakumar A, Hwang SJ, Fabiny JM, Chinault AC, Barnes EM (1989) Isolation of an ammonium or methylammonium ion transport mutant of *Escherichia coli* and complementation of the cloned gene. J Bacteriol 171:996-1001

Jayakumar A, Schulman L, MacNeil D, Barnes EM (1986) Role of the *Escherichia coli glnALG* operon in regulation of ammonium transport. J Bacteriol 166:281-284

Jouanneau Y, Meyer CM, Vignais PM (1983) Regulation of nitrogenase activity through iron protein interconversion into an active and an inactive form in *Rhodopseudomonas capsulata*. Biochim Biophys Acta 749:318-328

Kalb VF, Donohue TJ, Corrigan MG, Bernlohr RW (1978) A new and specific assay for ammonia and glutamine sensitive to 100 pmol. Anal Biochem 90:47-57

Kallas T, Dahlquist FW (1981) Phosphorus-31 nuclear magnetic resonance analysis of internal pH during photosynthesis in the cyanobacterium *Synechococcus*. Biochemistry 20:5900-5907

Kanemoto RH, Ludden PW (1984) Effect of ammonia, darkness, and methosulfate on whole-cell nitrogenase and Fe protein modification in *Rhodospirillum rubrum*. J Bacteriol 158:713-720

Kerby NW, Rowell P, Stewart WDP (1986) The uptake and metabolism of methylamine by N_2-fixing cyanobacteria. Arch Microbiol 143:353-358

Kerby NW, Rowell P, Stewart WDP (1987) Cyanobacterial ammonium transport, ammonium assimilation and nitrogenase regulation. New Zealand Mar Freshwater Res 21:447-456

Kleiner D (1981) The transport of NH_3 and NH_4^+ across biological membranes. Biochim Biophys Acta 639:41-52

Kleiner D (1985) Bacterial ammonium transport. FEMS Microbiol Rev 32:87-100

Krulwich TA, Guffanti AA (1989) Alkalophilic bacteria. Annu Rev Microbiol 433:435-463

Kurz WGW, LaRue TA, Chaston KB (1975) Nitrogenase in synchronized *Azotobacter vinelandii* OP. Can J Microbiol 21:984-988

Lattore C, Lee JH, Spiller H, Shanmugam KT (1986) Ammonium ion-excreting cyanobacterial mutants as a source of nitrogen for growth of rice: a feasibility study. Biotechnol Lett 8:507-512

Lea PJ, Joy KW, Ramos JL, Guerrero MG (1984) The action of 2-amino-4-(methylphosphinyl)-butanoic acid (phosphinothricin) and its 2-oxo-derivative on the metabolism of cyanobacteria and higher plants. Phytochem 23:1-6

Lea PJ, Mills WR, Wallsgrove RM, Miflin BJ (1982) Assimilation of nitrogen and synthesis of amino acids in chloroplasts and cyanobacteria (blue-green algae). In: Schiff JA (ed) On the origins of chloroplasts. Elsevier/North-Holland Inc, New York, pp 149-178

Martinez MR (1984) Algae: biofertilizer for rice. Philippines Council for Agriculture Research and Resources Development (PCARRD) Monitor 12:9-12

Meeks JC (1981) Radiotracer studies of nitrogen metabolism in cyanobacteria. In: Root JW, Krohn KA (eds) Advances in Chemistry, series 197. American Chemical Society, Washington, pp 269-294

Meeks JC, Steinberg N, Joseph CM, Enderlin CS, Jorgensen PA, Peters GA (1985) Assimilation of exogenous and dinitrogen-derived $^{13}NH_4^+$ by *Anabaena azollae* separated from *Azolla caroliniana* Willd. Arch Microbiol 142:229-233

Meeks JC, Wolk CP, Thomas J, Lockau W, Shaffer PW (1977) The pathways of assimilation of $^{13}NH_4^+$ by the cyanobacterium *Anabaena cylindrica*. J Biol Chem 252:7894-7900

Meeks JC, Wolk CP, Thomas J, Lockau W, Schilling N, Shaffer PW, Chien WS (1978) Pathways of assimilation of ^{13}N-N_2 and $^{13}NH_4^+$ by cyanobacteria with and without heterocysts. J Bacteriol 134:125-130

Murry MA, Benemann JR (1979) Nitrogenase regulation in *Anabaena cylindrica*. Plant Cell Physiol 20:1391-1401

Musgrave SC, Kerby NW, Codd GA, Stewart WDP (1982) Sustained ammonia production by immobilized filaments of the nitrogen- fixing cyanobacterium *Anabaena* 27893. Biotechnol Lett 8:507- 512

Newton JW, Cavins JF (1985) Liberation of ammonia during nitrogen fixation by a facultative heterotrophic cyanobacterium. Biochem Biophys Acta 809:44-50

Nierzwicki-Bauer SA, Haselkorn R (1986) Differences in mRNA levels in *Anabaena* living freely or in symbiotic association with *Azolla*. EMBO J 5: 29-35

Ohmori K, Ohmori M (1988) Ammonium assimilation in the blue- green alga *Spirulina platensis*. Japan J Phycol 36:12-16

Ohmori M, Hattori A (1978) Transient change in the ATP pool of *Anabaena cylindrica* associated with ammonia assimilation. Arch Microbiol 117:17-20

Ohmori M, Ohmori K, Strotmann H (1977) Inhibition of nitrate uptake by ammonia in a blue-green alga *Anabaena cylindrica*. Arch Microbiol 114:255-262

Papen H, Bothe H (1984) The activation of glutamine synthetase from the cyanobacterium *Anabaena cylindrica* by thioredoxin. FEMS Microbiol Lett 23:41-46

Paschinger H (1977) DCCD-induced sodium uptake by *Anacystis nidulans*. Arch Microbiol 113:285-291

Polukhina LE, Shakhurieva GN, Shestakov SV (1982) Ethylenediamine-resistant *Anabaena variabilis* mutants with derepressed nitrogen-fixing system. Microbiol 51:90-95

Querijero-Palacpac NM, Milagrosa M, Boussiba S (1990) Mass cultivation of the nitrogen-fixing cyanobacterium *Gloeotrichia natans*, indigenous to rice-fields. J Appl Phycol 2:319-325

Qian-Lin W, Yong Ding L, Yin-Wu S, Chuan-Yin J, Jin-Shu L, Jia-Min Z, Shang-Hao L (1991) Studies on mixed mass cultivation of *Anabaena* sp. on a large scale. Bioresources Technol 38:221- 228

Rai AK, Kashyap AK, Gupta SL (1981) ATP-dependent uptake of nitrate in *Nostoc muscorum* and inhibition by ammonium ions. Biochim Biophys Acta 674:78-86

Rai AK, Singh S (1982) Regulation of nitrate uptake in *Nostoc muscorum* by glutamine synthetase. FEMS Microbiol Lett 14:303-306

Rai AN, Rowell P, Stewart WDP (1984) Evidence for an ammonium transport system in free-living and symbiotic cyanobacteria. Arch Microbiol 137:241-246

Rai AN, Singh DT, Singh HN (1986) Regulation of ammonium/methylammonium transport by ammonium in the cyanobacterium *Anabaena variabilis*. Physiol Plant 68:320-322

Ramos JL, Guerrero MG, Losada M (1982) Photoproduction of ammonia from nitrate by *Anacystis nidulans*. Biochim Biophys Acta 679:323-330

Ramos JL, Guerrero MG, Losada M (1982) Sustained photoproduction of ammonia from nitrate by *Anacystis nidulans*. Appl Environ Microbiol 44:1020-1025

Ramos JL, Guerrero MG, Losada M (1983) Optimization of conditions for ammonia photoproduction by the nitrogen-fixing cyanobacterium *Anabaena* sp. strain ATCC 33047. In: Biotech'83, Proc International conference on the commercial applications and implications of biotechnology, Online Publisher, Ltd, Northwood, UK, pp 877-884

Ramos JL, Guerrero MG, Losada M (1984) Sustained photoproduction of ammonia from dinitrogen and water by the nitrogen-fixing cyanobacterium *Anabaena* sp. strain ATCC 33047. Appl Environ Microbiol 48:114-118

Ramos JL, Guerrero MG, Losada M (1987) Factors affecting the photoproduction of ammonia from dinitrogen and water by the cyanobacterium *Anabaena* sp. strain ATCC 33047. Biotechnol Bioengin 29:566-571

Reglinski A, Rowell P, Kerby NW, Stewart WDP (1989) Characterization of the methylammonium/ammonium transport system in mutant strains of *Anabaena variabilis* resistant to ammonium analogues. J Gen Microbiol 135:1441-1451

Reich S, Almon H, Böger P (1986) Short-term effect of ammonia on nitrogenase activity of *Anabaena variabilis* (ATCC 29413). FEMS Microbiol Lett 34:53-56

Ritchie, RJ (1987a) The permeability of ammonia, methylamine and ethylamine in the charophyte *Chara corallina (C. australis)* J Exp Bot 38: 67-76

Ritchie RJ, Gibson J (1987b) Permeability of ammonia and amines in *Rhodobacter sphaeroides* and *Bacillus firmus*. Arch Biochem Biophys 258:332-341

Ritchie RJ, Gibson J (1987) Permeability of ammonia, methylamine and ethylamine in the cyanobacterium *Synechococcus* R-2 (*Anacystis nidulans*) PCC 7942. J Membrane Biol 95:131-142

Roger PA, Kulasooriya SA (1980) Blue-Green Algae and Rice. The International Rice Research Institute, Los Banos, Philippines, pp 1-36

Roger PA, Santiago-Ardales S, Reddy PM, Watanabe I (1987) The abundance of heterocystous blue-green algae in rice soils and inocula used for application in rice fields. Bio Fert Soils 5:98-105

Roger PA, Watanabe I (1986) Techniques for utilizing biological nitrogen fixation in wetland rice: potentialities, current usage, and limiting factors. Fert Res 9:39-77

Ronzio RA, Meister A (1968) Phosphorylation of methionine sulfoximine by glutamine synthetase. Proc Natl Acad Sci USA 59:164-170

Schuldiner S, Rottenberg H, Avron M (1972) Determination of ΔpH in chloroplasts. Eur J Biochem 25:64-70

Singh DT, Modi DR, Singh HN (1986) Evidence for glutamine synthetase and methylammonium (ammonium) transport system as two distinct targets of methionine sulfoximine inhibitory action in the cyanobacterium *Anabaena doliolum*. FEMS Microbiol Lett 37:95-98

Singh DT, Rai AN, Singh HN (1985) Methylammonium (ammonium) uptake in a glutamine auxotroph of the cyanobacterium *Anabaena cycadae*. FEBS Letters 186:51-53

Spiller H, Lattore HC, Hassan ME, Shanmugam KT (1986) Isolation and characterization of nitrogenase-derepressed mutants of the cyanobacterium *Anabaena variabilis*. J Bacteriol 165:412-419

Stewart WDP (1980a) Some aspects of structure and function in N_2-fixing cyanobacteria. Annu Rev Microbiol 34:497-536

Stewart WDP (1980b) Transport and utilization of nitrogen sources by algae. In: Payne JW (ed) Microorganisms and nitrogen sources. John Wiley, New York, pp 577-607

Stewart WDP, Haystead A, Dharmawardene MWN (1975) Nitrogen assimilation and metabolism in blue-green algae. In: Stewart WDP (ed) Nitrogen fixation by free-living microorganisms. Cambridge University Press, Cambridge, pp 129-158

Stewart WDP, Rowell P (1975) Effects of L-methionine-DL-sulphoximine on the assimilation of newly fixed NH_3, acetylene reduction and heterocyst production in *Anabaena cylindrica*. Biochem Biophys Res Commun 65:846-856

Subrahmanian G, Shanmugasundaram S (1986) Uninduced ammonia release by the nitrogen fixing cyanobacterium *Anabaena*. FEMS Microbiol Lett 37:151-154

Thomas J, Meeks JC, Wolk CP, Shaffer PW, Austin SM, Chien WS (1977) Formation of glutamine from [^{13}N]ammonia, [^{13}N]dinitrogen, and [^{14}C]glutamate by heterocysts isolated from *Anabaena cylindrica*. J Bacteriol 129:1545-1555

Thomas J, Wolk CP, Shaffer PW (1975) The initial organic products of fixation of ^{13}N-labeled nitrogen gas by the blue- green alga *Anabaena cylindrica*. Biochem Biophys Res Commun 67:501-507

Thomas SP, Zaritsky A, Boussiba S (1990) Ammonium excretion by an L-methionine-DL-sulfoximine-resistant mutant of the rice field cyanobacterium *Anabaena siamensis*. Appl Environ Microbiol 56:3499-3504

Tubb RS, Postgate JR (1973) Control of nitrogenase synthesis in *Klebsiella pneumoniae*. J Gen Microbiol 79:103-117

Turpin DH, Edie SA, Canvin DT (1984) In vivo nitrogenase regulation by ammonium and methylamine and the effect of L-methionine-DL-sulfoximine on ammonium transport in *Anabaena flos-aquae*. Plant Physiol. 74:701-704

Venkataraman GS (1975) The role of blue green algae in tropical cultivation. In: Stewart WDP (ed) Nitrogen fixation by free-living microorganisms. Cambridge Univ Press, London, pp 207- 218

Venkataraman LV (1986) Blue-green algae as biofertilizers. In: Richmond A (ed) CRC Handbook of Microalgal Mass Culture, CRC Press, Boca Raton, Florida, pp 455-471

Vonshak A, Richmond A (1988) Mass production of the blue-green algae *Spirulina*: an overview. Biomass 15:233-247

Watanabe I (1984) Use of symbiotic and free-living blue-green algae in rice culture. Outlook Agric 13:166-172

Wolk CP (1980) Heterocysts, ^{13}N, and N_2-fixing plants. In: Newton WE, Orme-Johnson WH (eds) Nitrogen fixation, vol II. University Park Press, Baltimore, pp 279-292

Wolk CP, Thomas J, Shaffer PW, Austin SM, Galonsky A (1976) Pathway of nitrogen metabolism after fixation of ^{13}N-labeled nitrogen gas by the cyanobacterium *Anabaena cylindrica*. J Biol Chem 251:5027-5034

Yoch DC, Zhang Z, Claybrook DL (1983) Methylamine metabolism and its role in nitrogenase "switch off" in *Rhodopseudomonas capsulata*. Arch Microbiol 134:45-48

Zimmerman WJ, Boussiba S (1986) Ammonia assimilation and excretion in an asymbiotic strain of *Anabaena azollae* from *Azolla filiculoides* Lam. J Plant Physiol 127:443-450

Zumft WG, Castillo F (1978) Regulatory properties of the nitrogenase from *Rhodospirillum palustris*. Arch Microbiol 117:53-60

Urea Metabolism

ASHWANI K. RAI

Department of Botany, Banaras Hindu University, Varanasi 221005

1 Introduction

Cyanobacteria are probably one of the most important group of organisms of serious global ecological importance as they populate many different environments. They are common constituents of the microbial flora of forest and agricultural soils, specially of tropical paddy fields and waters rich in organic matter. The composition and size of cyanobacterial flora in these habitats are controlled by many interacting factors like temperature, light, nutrient concentrations, growth-promoting and growth-inhibiting substances and selective grazing (FOGG 1965, AHLGREN 1970). Among them, nutrient factor is the one most influenced by man to any great extent and may lead to change in the cyanobacterial species composition. Which of the nutrient is required in largest quantity is based on the elemental analysis of cells. Elements like C, H and O taken into the cells as H_2O and gaseous CO_2, are energetically inexpensive in terms of transport and often not limiting. However, except for N_2-fixing cyanobacteria, the total nitrogen in the cyanobacterial cells is a measure of nitrate, ammonia, urea or any other inorganic or organic forms of nitrogen available. Nitrate and ammonia have received much attention, but, a little is known about urea utilization.

Urea is the first organic molecule to be synthesized in laboratory (WÖHLER 1828) and the enzyme urease the first enzyme to be crystallized (SUMNER 1926). Both the discoveries revolutionized the theories on the origin of organic compounds and the nature (proteinous) of enzymes, respectively. A good deal of knowledge has been achieved on different aspects of urea assimilation; on the physiology, biochemistry and genetics of the process as it takes place in eukaryotic organisms and eubacteria. Unfortunately, there has not been, however a parallel development in the field that concerns the cyanobacteria. It is likely that a group as large and as ancient as cyanobacteria would possess diversity in biochemical pathways or regulatory mechanisms or both. Hence, present review is a first time attempt to summarize the available informations which concern urea utilization by cyanobacteria, ecological significance, urea uptake and its subsequent hydrolysis, the factors regulating the processes, physico-chemical properties of the cyanobacterial urease and its comparison with some well studied eukaryotic and microbial ureases.

2 Ecology

2.1 Environmental urea

Urea is continuously fed into the environment through biological activity. Most terrestrial animals (all the mammals including human), adult amphibians and cartilaginous fishes excrete amino nitrogen in the form of urea. Urea measures 0.4 to 0.5 M in the urine of human and accounts 10 kg of urea released per adult per year (GRIFFITH et al. 1976). Uric acid, the excreted end product of purine catabolism in primates, birds, reptiles and insects is degraded to urea. In teleost fishes and other vertebrates uric acid is degraded further to the excretory product allantoate and allantoin, respectively. Degradation of purines and pyrimidines leads to uric acid and urea, respectively. Guanosine 5′-monophosphate catabolism also yields uric acid as end product. Uric acid excretion by an adult human is about 0.6 g per day arising from ingested purines and from turnover of purines nucleotides of nucleic acids. The nitrogenous compounds arginine, agmatine, allantoate and allantoin, all release urea after their degradation. This urea is further metabolized by plants and microorganisms. Hence, in addition to serving the needs of organisms for nitrogen, urea breakdown is important to the global nitrogen cycle.

2.2 Urea as fertilizer

Agricultural productivity is largely determined by the availability of fixed nitrogen, added as fertilizer or introduced by the microbial nitrogen fixation. The major role played by N_2-fixing cyanobacteria in restoring nitrogen to the tropical paddy fields contributing to the soil fertility has been well documented (SINGH 1961).

Urea is the most common source of nitrogenous fertilizer used in tropics. Its popularity is because of its high nitrogen content (47% N), low unit cost, ease in handling and ready availability in the world market. However, efficiency of the urea fertilizer depends upon its controlled hydrolysis. From surface applied urea, a major portion of nitrogen escapes into the atmosphere as NH_3. This is one of the primary reasons for its inefficiency (FENN and HOSSNER 1985). WATSON et al. (1994a) reported volatilization of 39% of applied urea as NH_3, while MULVANEY and BREMNER (1981) noted a loss of 50%. NH_3 volatilization was maximum in soils with high pH values. Damage to the seedlings and early plant growth by urea fertilizer is of common occurrence (BREMNER and KROGEIER 1988). Ammonia toxicity and elevated pH are the primary reasons for this adverse effect. The system which can reduce the rate and/or delay urea hydrolysis will promote its diffusion into the soil, due to which build up of NH_3 concentration on the localized soil surface (conducive of NH_3 volatilization) is avoided, and hence damage to the plant is prevented. This may be accomplished by altering granule size, using chemical additives, slow release systems and urease inhibitors (GOULD et al. 1986). The later in this context appears to be the most promising (BREMNER and CHAI 1989). However, in the selection of inhibitors, it should be kept in mind that they have long-lasting effect, identical solubility and diffusibility to urea as well as non-toxic. WATSON and associates in a series of field experiments (WATSON et al. 1990, 1994a,b) found that urea amended with 0.5% nBTPT (N-(n-butyl) thiophosphoric triamide) reduces ammonia volatilization up to 97% depending upon the soil quality, increases N-recovery and dry matter yield of grass by 9% and the total N-recovery in soil-plant system by 17%, compared to urea alone. These observations are for the temperate region. Hence, needs are there to study many other different inhibitors suitable for tropics.

2.3 Effect on Cyanobacterial flora

Due to an increased and widespread use of N-fertilizers, it is imperative to consider the effect of such fertilizers on the composition and size of the cyanobacterial population, nitrogen-fixing ability and the resultant role in N-cycling in the crop field habitats. The abundant growth of algae and cyanobacteria therein suggests that they might use a large portion of fertilizers applied to the fields, thereby reducing the availability of nitrogen to crop plants. It is also not known whether the nitrogen-fixing cyanobacteria are eliminated from the field or such species have the capacity to utilize these sources and establish themselves in a significant population contributing to the nitrogen input in the field.

Since nitrogen is often the limiting nutrient for aquatic algae and cyanobacteria unable to use dinitrogen (WETZEL 1975), N_2-fixing cyanobacteria under N-deficient conditions should generally be favoured and due to lack of competitiveness of the other algae, can develop profusely if the other environmental factors are not limiting.

Under N-rich conditions, as diazotrophic organisms, they can use mineral nitrogen for their growth, but at the same time they have to compete with non-N_2-fixing cyanobacteria and eukaryotic algae.

Only a few studies have been conducted on the competition between N_2-fixing and non-N_2-fixing forms as affected by the nature and concentration of combined nitrogen (ROGER and REYNAUD 1979) and the results have been contradictory. RINAUDO (1974) reported the inhibitory effect of nitrogen fertilizers on N_2-fixing cyanobacteria. THAN-TUN (1969) observed that nitrogenous fertilizers increased the total algal biomass and depressed the growth of *Anabaena* and that in soil treated with ammonium sulphate or calcium cyanide, only green algae appeared dominant. OKUDA and YAMAGUCHI (1952) reported N_2-fixing forms to become abundant only in the unfertilized soils. However, MAHAPATRA et al. (1971) reported that addition of N to soils induced the dominance of *Nostoc muscorum, Anabaena cylindrica* and *Volvox* sp. With regard to the concentration of combined nitrogen present in the environment, WARD and WETZEL (1980) found that growth rate of N_2-fixing cyanobacteria varied according to the amount of combined nitrogen present in the medium. In the field conditions, it is difficult to determine as to what extent, the various nitrogenous forms are involved in the composition of cyanobacterial and algal flora. There also exists a considerable variation among different cyanobacterial and algal species as to which forms of nitrogen can be assimilated preferentially (TILMAN 1977, RAI and RAI 1977, TURPIN and HARRISON 1979). The nature of nitrogen source available influences the flora in addition to producing different physiological response and alteration in cellular composition (EPPLEY et al 1969, TILMAN 1976, KILHAM and KILHAM 1978, SINGH and RAI 1989a).

3 Urea transport

Entry of urea into the cells is the first step in urea metabolism. Despite information available on the ability of cyanobacteria to utilize urea as a sole source of nitrogen, there are meagre reports on the kinetics of urea uptake by cyanobacteria. However, observations on urea uptake by green algae (WILLIAMS and HODSON 1977, REES and SYRETT 1979), fungi (PATEMAN et al. 1982) and bacteria (JAHNS et al. 1988) have provided evidence for the carrier mediated urea uptake. Experiments under different metabolic conditions have indicated that the energy for the uptake of urea is derived mainly from cyclic photophosphorylation.

Urea transporter of *Anabaena doliolum* and *Anacystis nidulans* possessed a K_m of 250 and 400 μM urea, respectively at optimum pH (7.5) and temperature (30°C) (RAI and SINGH 1987a). The observed affinity for urea was however, lower than that reported for bacteria and fungi (MOBLEY and HAUSINGER 1989). Uptake of urea exhibited two phenotypes (1) repressed state (cells grown with ammonia) and (2) constitutive state (growth with N_2 and nitrate). Stimulation of urea uptake in N-starved cells may be presumed for the increased porter density in the membrane. Presence of ammonia in the medium did not inhibit the uptake of urea but simultaneous presence of urea depressed ammonia uptake (HEALEY 1977, RAI and SINGH 1987a). Since intracellular ammonia inhibits its own uptake, it is assumed that urea taken up by the cells is hydrolyzed to ammonia and in turn represses its own uptake.

4 Enzymology of urease

The degradation of urea in microorganisms is affected by two enzymes. One is urease (urea amidohydrolase [EC 3.5.1.5]) which is widely observed among bacteria, yeast, fungi and higher plants and catalyzes the hydrolysis of urea into ammonium carbamate:

$$CO(NH_2)_2 + 2H_2O \longrightarrow (NH_4)_2CO_3$$

Ammonium carbamate thus formed, in a nonenzymatic reaction and in the presence of water dissociates into the final product ammonium and bicarbonate ions.

The other mechanism for urea degradation involves ATP:urea amidolyase and is present in many fungi and chlorophycean algae (ROON and LEVENBERG 1968, BEKHEET and SYRETT 1977). This ATP- and biotin-dependent enzyme system is separated into two components (THOMPSON and MUENSTER 1971): Enzyme one is urea carboxylase [urea:CO_2-ligase (ADP), EC 6.3.4.6] catalyzing the reaction:

$$CO(NH_2)_2 + HCO_3^- + ATP \underset{\text{Biotin}}{\overset{Mg^{+2}\,K^+}{\rightleftharpoons}} \text{allophanate} + ADP + Pi$$

This allophanate produced is then used as substrate for the second enzyme, allophanate amidohydrolase [EC 3.5.1.13]

$$\text{allophanate} + 2H_2O + OH^- \longrightarrow 2HCO_3^- + 2NH_3$$

The organisms have either of them, urea amidolyase or urease, no organism contains both (LEFTLEY and SYRETT 1973, SYRETT and LEFTLEY 1976). These enzymes can be distinguished from each other (ROON and LEVENBERG 1972): ATP:urea amidolyase requires ATP and biotin for its activity and is inhibited by avidin which combines with the latter, while urease activity is inhibited by hydroxyurea.

Enzyme extracts from cyanobacterial species *Phormidium luridum*, *Plectonema calothricoides* (BERNS et al. 1966), *N. muscorum* (ROON and LEVENBERG 1972), *Aphanocapsa* 6308 (WEATHERS et al. 1978), *Spirulina maxima* (CARVAJAL et al. 1982), *A. cylindrica* (MACKERRAS and SMITH 1986) *A. doliolum*, *A. nidulans* (RAI and SINGH 1987b) and *Anabaena variabilis* (GE et al. 1990) when tested, revealed urease as the urea hydrolyzing enzyme.

4.1 Kinetics

Kinetics of cyanobacterial urease have been worked out at given temperature, pH values, ionic strength and in specific buffer systems usually at low product concentration. *A. Cylindrica* crude cell-free urease showed a K_m value of 1.3 mM in HEPES buffer, pH 7.5 at 30°C (MACKERRAS and SMITH 1986). CARVAJAL et al. (1982) found a K_m value of 0.12 mM for purified *S. maxima* urease in tris-maleate buffer, pH 8.7 at 37°C. Urease, purified from *A. doliolum* exhibited a K_m of 0.115 mM at optimum pH of 7.0 and temperature 40°C in phosphate buffer (RAI 1989), similar to that of *S. maxima* (CARVAJAL et al. 1982). *A. Variabilis* crude cell-free urease showed a K_m of 46.2 μM in tris-HCl buffer pH 7.0 at 37°C (GE et al. 1990). The data on the K_m value of purified urease from *A. doliolum* (0.115 mM) (RAI 1989) agree with the value observed for crude cell-free extracts (0.12 mM) (RAI and SINGH 1987b) facilitating the K_m determination even from crude ureases. All the K_m values mentioned above are for filamentous cyanobacterial ureases. The unicellular non-nitrogen fixing cyanobacterium *A. nidulans,* however exhibited biphasic kinetics, K_m of 0.25 and 1.66 mM (RAI and SINGH 1987b). At 5 M external urea (preferred concentration for the growth of *A. nidulans*), the low affinity system would account for a significant portion of the total urea hydrolysis indicating that the cyanobacterium has an inbuilt mechanism to avoid high intracellular ammonium generation from urea hydrolysis under high concentrations of urea.

While comparing the K_m values, it should be kept in mind that K_m varies with buffer and pH conditions; buffer ions and their higher ionic strength decrease the activity of urease, thus Michaelis constant increases with the addition of buffers and salts (FASMAN and NIEMANN 1951, KISTIAKOWSKY et al. 1952). Urease in a buffer-free solution is less susceptible to the inhibition of substrate product (QIN and CABRAL 1994). The cyanobacterial ureases in general possess significantly lower K_m values than that of jack bean and other microbial ureases (for comparison refer MOBLEY and HAUSINGER 1989) making cyanobacterial urease preferable over ureases from other sources in detecting urea.

The specific activity of crude urease from *A. cylindrica* was in the range of 16-58 nmol min^{-1} mg $protein^{-1}$. Cyanobacterial urease purified from *S. maxima* (CARVAJAL et al. 1982) and *A. doliolum* (RAI 1989) exhibited significantly lower activities (9.27 and 9.39 units mg^{-1} protein, respectively) than that of many microbial ureases (MOBLEY and HAUSINGER 1989) and well characterized jack bean urease (ANDREWS et al. 1984). May be, the cyanobacterial ureases have actually lower activity or alternatively the enzymes are inactivated during purification. Since specific activity of any enzyme depends upon its purification state, the cyanobacterial ureases have not been purified completely. Thus, before reaching any conclusion it is desirable to test other cyanobacterial strains.

4.2 Structural properties

Cyanobacterial urease remains little explored from structural view point with only report on the enzyme purification from a halophilic, nonheterocystous cyanobacterium *S. maxima* (CARVAJAL et al. 1982) and freshwater, heterocystous cyanobacterium *A. doliolum* (RAI 1989). Molecular weight of the native enzyme measured by gel filtration through Sephadex G-200 column is approximately 230,000 and consists of six subunits of identical size with a *Mr* of 37,000. Bacterial ureases lie in the same range (*Mr* = 200,000 to 250,000), but happens to be significantly smaller than the eukaryotic jack bean urease (*Mr* = 590,000). Jack bean urease is hexamer of identical subunits with *Mr* of 95,000 (DIXON et al. 1980) as found in cyanobacteria. However, there always existed a discrepancy in the number of bacterial ureases subunits. MOBLEY and HAUSINGER (1989) have generalized that gram-positive bacteria possess homopolymeric urease, whereas gram-negatives have three distinct subunit types.

4.2.1 Ni

Ni is an essential component of urease (FISHBEIN et al. 1976) as shown in plants and microorganisms (MOBLEY and HAUSINGER 1989) and is required for the synthesis of urease in many organisms including cyanobacteria (MACKERRAS and SMITH 1986). Although Ni is not included in the list of micronutrients for the growth of cyanobacteria, stimulates the growth of *A. doliolum* and *A. nidulans* at a concentration of 0.01 μM (SINGH and RAI 1990). Some algal species could not grow on urea or utilize it poorly in the Ni-deficient environment (MCCARTHY 1972, ANTIA et al. 1975, OLIVEIRA and ANTIA 1984). Moreover, VAN BAALEN and O'DONNELL (1978) isolated a species of *Oscillatoria* exhibiting an absolute requirement of Ni.

The level of urease activity depended on the concentration of Ni in the growth medium (MACKERRAS and SMITH 1986). Addition of Ni to cultures grown under Ni-deficient condition resulted in the recovery of activity even when protein synthesis was halted (by incubating the cells with chloramphenicol). This indicated that Ni is not needed for the synthesis of urease apoenzyme which can be activated by subsequent addition of Ni. Specific inhibitors, reported for bacterial and plant ureases hydroxyurea, acetohydroxamic acid and phenyl-phosphorodiamidate inhibit cyanobacterial urease revealing the presence of Ni co-factor. Tight bond of Ni ion to the protein is evidenced by the retention of the enzyme activity during isolation in buffer containing 1 mM EDTA (RAI 1989). Although sufficient evidences are there in favour of Ni requirement for urease activity, it still needs direct confirmation in the case of cyanobacteria.

4.3 Inhibitors

Urease inhibitors, helpful in unfolding the enzyme action mechanism are the most promising tools to improve the efficiency of urea fertilizers. In general, urease inhibitors are the substrate analogues such as thiourea, hydroxyurea and hydroxamic acids. Acetohydroxamic acid which has been used on several microorganisms and on plant ureases is a good metal chelator and is thought to bind to the active site nickel ion. It is a reversible, competitive inhibitor of urease with slow dissociation rate (ANDREWS et al. 1984). Phosphoromides (phenylphosphorodiamidate, N-acylphosphoric triamide and N-(n-butyl) thiophosphoric triamide) are the potent urease inhibitors and act as transition state analogue and bind to the active site nickel. Fully protonated phosphoric acid, thiols, boric acids, boronic acids, alkylating agents (*N*-ethylmaleimide, iodoacetamide and iodoacetic acid), reagents reactive to thiol group (*p*-chloromercuribenzoate) and disulfide reagent (5,5-dithiobis (2-nitrobenzoic acid) are reported to inactivate ureases (MOBLEY and HAUSINGER 1989). On cyanobacterial urease sulfhydryl reagent *p*-hydroxymercuribenzoate, acetohydroxamic acid and phenylphosphorodiamidate including hydroxyurea have been tested and were found inhibitory to the enzyme (RAI and SINGH 1987b, RAI 1989, GE et al. 1990).

4.4 Stability

In general, urease is highly sensitive to many foreign chemical and heavy metals. Because of this property, it is proposed to be an useful indicator to determine the presence in the environment of trace amounts of heavy metal ions (OLSON and CHRISTENSEN 1982). Ge et al. (1990) observed the crude urease from *A. variabilis* relatively heat resistant, which retained most of the activity at 30°C for several hours and 25% of activity after 2h incubation at 50°C. The *Tm* for denaturation was 73°C. Activity of purified urease from *A. doliolum* declined by 70% over two weeks period when stored at 4°C in 50 mM phosphate buffer, pH 7.0 containing 1 mM EDTA and 5 mM β-mercaptoethanol, but lowering the temperature to -20°C and addition of 20% glycerol in the buffer made the enzyme quite stable and almost no loss of activity was observed over two months of storage (RAI 1989).

4.5 Cellular localization

Contrary reports exist on the localization of urease in different organisms. Some studies (BOOTH and VISHNIAE 1987) suggest the enzyme as an extracellular, while others (McLEAN et al. 1985) describe the enzyme attached to periplasm and outer membrane. However, majority of the studies (MOBLEY and HAUSINGER 1989) found it a soluble protein located in the cytoplasm in bacteria, fungi and higher plants. We found that most of the urease activity of the cyanobacterium *A. doliolum* is present in spheroplast lysate as a soluble enzyme which is released from glycerol protected spheroplasts after osmotic shock and lysis (RAI and SINGH 1987b, SINGH and RAI 1989b). Immunogold localization using urease antibodies may give a final answer to the precise localization of urease.

4.6 Regulation

In cyanobacteria, assimilation of nitrogen is under strict regulation. Ammonia, a product of urea hydrolysis when available represses the synthesis of proteins involved in the assimilation of inorganic nitrogen sources alternative to ammonia. It is proposed that *ntcA* gene exerts positive control on the nitrogen regulated genes by binding directly to the promoters of the nitrogen regulated genes and consequently affects the mRNA levels (LUQUE et al. 1994).

Urea may serve as a source of nitrogen for most of the cyanobacteria and its presence in the nutrient solution represses nitrogenase synthesis (RAWSON 1985) at the transcription level by the same mechanism as ammonia (GE et al. 1990). Cyanobacterial ureases so far reported are constitutive and does not require urea for its synthesis (MACKERRAS and SMITH 1986, RAI and SINGH 1987b). Ammonia does not repress urease synthesis in *A. cylindrica* (MACKERRAS and SMITH 1986). GE et al. (1990) observed about 50% higher specific activity when *A. variabilis* cells were grown with ammonia, while ureases of *A. doliolum* and *A. nidulans* were partially repressed during growth with ammonia (RAI and SINGH 1987b). It is possible that the observed differences are due to growth conditions, period of storage of the extracts or species specific.

On the other hand, while urease is always constitutive in algae (SYRETT and LEFTLEY 1976, BEKHEET and SYRETT 1977), the enzyme requires urea as an inducer in bacteria such as *Proteus mirabilis* and *Providencia rettgeri* (ROSENSTEIN et al. 1980). Ammonia represses the urease synthesis in *Hydrogenomonas* spp. *Bacillus megaterium, Micrococcus cerificans* (KRAMER et al. 1967, KALTWASSER et al. 1972) *Klebsiella aerogenes* (FRIEDRICH and MAGASANIK 1977) and *Bifidobacterium* (CROCIANI and MATTEUZZI 1982), while in *Klebsiella aerogenes* and *Pseudomonas aeruginosa* it is found to be derepressed under nitrogen-limiting conditions (JANSSEN et al. 1981). In *Proteus vulgaris* and *Sporosarcina ureae*, the enzyme is formed constitutively (KRAMER et al. 1967). Other forms like lichen *Peltigera canina* (having *Nostoc* cyanobiont) and *Peltigera aphthosa* (with both *Nostoc* as cyanobiont and *Coccomyxa* as phycobiont) possess the constitutive urease (SHAPIRO 1977, BROWN et al. 1982). At the same time, urease is inducible in lichens *Cladonia* species, *Parmelia roystonea, Catraria islandica, Hypogymnia physodes, Lobaria pulmonaria* and *Evernia prunastri* which have green algae as phycobiont (SHAPIRO 1977, VICENTE et al. 1978, XAVIER FILHO and VICENTE 1978, VICENTE and XAVIER FILHO 1979, CIFUENTES et al. 1981, LEGAZ and BROWN 1983). It seems that cyanolichen's urease is constitutive, whereas lichens with phycobionts (green algae) have inducible urease. As of now, the data on cyanobacterial urease are a few and in the light of the above findings, study of several cyanobacterial species are needed to establish urease regulation.

5 Role of urease

Urease, undoubtedly plays an important role when exogenous source of nitrogen is urea, but the fact that when urea is not the source of nitrogen for cyanobacterial growth, the enzyme is constitutive. This is suggestive of some additional function associated with the enzyme. The role of urease in cyanobacterial nitrogen metabolism has not yet been established. However, a general view is that urease plays no major role because its substrate, urea is not a major plant metabolite. POLACCO and HOLLAND (1993) have observed that considerable amount of plant nitrogen flow through urea which can be recycled only by urease, as well as abundant seed urease may play a chemical defense role by inducing a hostile environment upon microbial and insect attack.

GE et al. (1990) speculated cyanobacterial urease a scavenger of amino nitrogen from urea produced internally by degradation of arginine and purines. Cyanobacterial cells possess cyanophycin granules which consists of multi-L-arginyl-poly (L-aspartic acid) (SIMON 1971, ALLEN et al. 1980, ALLEN and WEATHERS 1980) as a source of nitrogen reserve and is mobilized during nitrogen starvation. Its degradation is accomplished by cyanophycinase to aspartic acid-arginine dipeptide (GUPTA and CARR 1981a). Free arginine is probably the product of the cleavage of arginine- aspartic acid dipeptide, which is further degraded to urea. Since cyanobacteria possess arginase activity (GUPTA and CARR 1981b), it may indicate the presence of urea producing pathway in cyanobacteria. In support lies the observations of GE et al. (1990), who have noted a significant stimulation of crude cell-free urease by arginine.

6 Method of estimation

Methods for the assay of urease activity fall into either of the categories: (1) direct assay of urea, (2) those dependent upon ammonia, (3) CO_2 release and (4) increase in pH of the reaction mixture. A decent review of methods for determination of urea nitrogen is given by MARTINEK (1969). To begin with, MOBLEY and HAUSINGER (1989) recommend HEPES buffer (50 mM) with 1 mM EDTA, pH 7.5 in the study of urease activity.

Among the methods, Fearon reaction (FEARON 1939) is direct interaction of urea with diacetyl monoxime to produce pink chromogen, which is measured spectrophotometrically between 515 to 540 nm.

Diacetylmonoxime + urea ⟶ pink chromogen + hydroxylamine

A combination of ferric chloride and thiosemicarbazide is used to stabilize and intensify the colour (WOOTTON 1974). The method is simple, rapid and highly sensitive.

In phenol nitropruside colorimetric method (CHANEY and MARBACH 1962), ammonia formed due to urea hydrolysis reacts with alkaline hypochlorite and phenol in the presence of a catalyst sodium nitropruside, $[Na_2Fe(CN)_5\,NO.2H_2O]$, to form indophenol which is measured spectrophotometrically at 570 nm. The concentration of phenate and hypochlorite reagent has been modified by KAPLAN (1969). The method is highly sensitive and can detect < 0.02 µmol of ammonia.

Nesslerization method is another spectrophotometric method in which ammonia reacts with Nessler's salt to give an orange colour. Although the method is less sensitive but rapid.

$$\underset{\text{Nessler's salt (Yellow)}}{NH_4OH + 2(KI)_2\,HGI_2} \longrightarrow \underset{\text{Dimercuric ammonium iodide (Yellow orange)}}{NH_2HG_2I_2} + 5KI + 2H_2O$$

In enzymic assay, ammonia aminates 2-oxoglutaric acid to glutamic acid in the presence of glutamate dehydrogenase (GDH) with the concurrent oxidation of NAD(P)H to NAD(P), which can be measured at 340 nm.

$$NH_3 + \text{2-Oxogluterate} + NADH \xrightarrow{GDH} \text{glutamic acid} + H_2O + NAD$$

106 cm^2/mol (ε 340=6.2) ε 340=0)

The method can detect 0.02 µmol of ammonia, but the disadvantage is the difference in pH optimum (pH 8.3) for glutamate dehydrogenase and urease.

Ammonium ion sensitive electrodes can also be used to measure ammonium ions, but with low sensitivity (0.1 mM). However, ammonia sensitive electrodes have higher sensitivity than the former (10 nmol of ammonia).

Increase in pH due to urea hydrolysis can be measured in the presence of pH indicator phenol red. This method is quite simple and useful to detect urease activity in native gel. The gel is equilibrated in a pH buffer 6.0, containing pH indicator then transferred to the solution containing urea. As urea is hydrolyzed pH indicator changes the gel colour at the location of urease activity which can be fixed by lead acetate (BLATTLER et al. 1967). Another method to stain urease electrophoresed on acrylamide gels is p-nitro blue tetrazolium salt. The gel is first equilibrated for 1-2 h with 0.05 M citrate buffer, pH 6.0, then transferred into the staining solution (38 ml water, 3.6 ml of 50 mM citrate buffer, pH 6.0, 5.2 ml M urea, 5 ml 0.5% p-nitro blue tetrazolium and 14 ml dithiothreitol). The reaction is stopped by N HCl.

$^{14}CO_2$ release from [^{14}C] urea is widely used to assay urease activity. [$^{14}CO_2$] is trapped in a tube containing phenethylamine-methanol (1:1, v/v). A portion of the trapping solution is counted for $^{14}CO_2$ by scintillation counting and gives the amount of $^{14}CO_2$ produced or the amount of urea hydrolyzed.

References

Ahlgren G (1970) Limnological studies of lake Norrviken, a eutrophicated Swedish lake. II. Phytoplankton and its production. Schweiz Z Hydrol 32:354-396

Allen MM, Hutchison F, Weathers PJ (1980) Cyanophycin granule polypeptide formation and degradation in the cyanobacterium *Aphanocapsa* 6308. J Bacteriol 141:687-693

Allen MM, Weathers PJ (1980) Structure and composition of cyanophycin granules in the cyanobacterium *Aphanocapsa* 6308. J Bacteriol 141:959-962

Andrews RK, Blakeley RL, Zerner B (1984) Urea and urease. In: Eichhorn GL, Marzilli LG (eds) Advances in inorganic biochemistry, vol 6. Elsevier Science Publishing Inc, New York, pp 245-283

Antia NJ, Berland BR, Bonin DJ, Maestrini SY (1975) Comparative evaluation of certain organic and inorganic sources of nitrogen for phototrophic growth of marine microalgae. J Mar Biol Ass UK 55:519-539

Bekheet IA, Syrett PJ (1977) Urea degrading enzymes in algae. Br Phycol J 12:137-143

Berns DS, Holohan P, Scott E (1966) Urease activity in blue-green algae. Science 152:1077-1078

Blattler DP, Contaxis CC, Reithel FJ (1967) Dissociation of urease by glycol and glycerol. Nature 216:274-275

Booth JL, Vishniae HS (1987) Urease testing and yeast taxonomy. Can J Microbiol 33: 396-404

Bremner JM, Chai HS (1989) Effects of phosphoroamides on ammonia volatilization and nitrite accumulation in soils treated with urea. Biol Fertil Soils 8:227-230

Bremner JM, Krogeier MJ (1988) Estimation of the adverse effects of urea fertilizer on seed germination, seedling growth and early plant growth. Proc Natl Acad Sci USA 85:4601-4604

Brown DH, Vicente C, Legaz E (1982) Urease activity in *Peltigera canina* (L) Willd. Crypto Bryol Lichenol 3:33-38

Carvajal N, Fernández M, Rodríguez JP, Donoso M (1982) Urease of *Spirulina maxima*. Phytochem 21:2821-2823

Chaney AL, Marbach EP (1962) Modified reagents for determination of urea and ammonia. Clinical Chem 8:130-132

Cifuentes B, Estvez MP, Vicente C (1981) *In vivo* protection of urease of *Evernia prunastri* by dithiothreitol. Physiol Plant 53:245-248

Crociani F, Matteuzzi D (1982) Urease activity in the genus *Bifidobacterium*. Ann Microbiol 133:417-423

Dixon NE, Hinds JA, Fihelly AK, Gazzola C, Winzor DJ, Blakeley RL, Zerner B (1980) Jack bean urease [EC 3.5.1.5]. IV. The molecular size and the mechanism of inhibition by hydroxamic acids. Spectrophotometric titration of enzymes with reversible inhibitors. Can J Biochem 58:1323-1334

Eppley RW, Rogers JN, McCarthy JJ (1969) Half-saturation constants for uptake of nitrate and ammonium by marine phytoplankton. Limnol Oceanogr 14:912-920

Fasman GD, Niemann C (1951) A reinvestigation of the kinetics of the urease-catalyzed hydrolysis of urea. I. The activity of urease in the presence of sodium and potassium phosphate. J Am Chem Soc 73:1646-1650

Fearon WR (1939) The carbamide biacetyl reaction: a test for citrulline. Biochem J 33:902-907

Fenn LB, Hossner LR (1985) Ammonia volatilization from ammonium or ammonium-forming nitrogen fertilizers. Adv Soil Sci 1:123-169

Fishbein WN, Smith MJ, Nagarajan K, Scurzi W (1976) The first natural nickel metalloenzyme: Urease. Fed Proc Fed Am Soc Exp Biol 35: 1680-1687

Fogg GE (ed) (1965) Algal culture and phytoplankton ecology. University of Wisconsin Press, Madison Milwaukee

Friedrich B, Magasanik B (1977) Urease of *Klebsiella aerogenes*: control of its synthesis by glutamine synthetase. J Bacteriol 131:446-452

Ge X, Cain K, Hirschberg R (1990) Urea metabolism and urease regulation in the cyanobacterium *Anabaena variabilis*. Can J Microbiol 36:218-222

Gould WD, Hagedorn C, McCready RGL (1986) Urea transformations and fertilizer efficiency in soil. Adv Agron 40:209-238

Griffith DP, Musher DP, Itin C (1976) Urease. The primary cause of infection-induced urinary stones. Invest Urol 13:346-350

Gupta M, Carr NG (1981a) Enzyme activities related to cyanophycin metabolism in heterocysts and vegetative cells of *Anabaena* sp. J Gen Microbiol 125:17-23

Gupta M, Carr NG (1981b) Enzymology of arginine metabolism in heterocyst-forming cyanobacteria. FEMS Microbiol Lett 12:179-181

Healey FP (1977) Ammonium and urea uptake by some freshwater algae. Can J Bot 55:61-69

Jahns T, Zobel A, Kleiner D, Kaltwasser H (1988) Evidence for carrier-mediated, energy dependent uptake of urea in some bacteria. Arch Microbiol 149:377-383

Janssen DB, Habets WJA, Marugg JT, van der Drift C (1982) Nitrogen control in *Pseudomonas aeruginosa*: mutants affected in the synthesis of glutamine synthetase, urease and NADP-dependent glutamate dehydrogenase. J Bacteriol 151: 22-28

Kaltwasser H, Kramer J, Conger WR (1972) control of urease formation in certain aerobic bacteria. Arch Mikrobiol 81:178-196

Kaplan A (1969) The determination of urea, ammonia and urease. In: Glick D (ed) Methods of biochemical analysis, vol 17. John Wiley & Sons, New York, pp 311-324

Kilham SS and Kilham P (1978) Natural community bioassays: predictions of results based on nutrient physiology and competition. Vern Int Ver Limnol 20:68-74

Kistiakowsky GB, Mangelsdorf PC, Rosenberg A Jr, Shore WHR (1952) The effects of electrolytes on urease activity. J Am Chem Soc 74:5015-5020

Kramer J, Kaltwasser H, Schlegel HG (1967) Die Bedeutung der urease repression fr die taxonomische klassifizierung von Bakterien. Zentralbl Bakteriol Parasitenkd Infektionskr Hyg Abt 2, 121:414-423

Leftley JW, Syrett PJ (1973) Urease and ATP:Urea amidolyase activity in unicellular algae. J Gen Microbiol 77:109-115

Legaz ME, Brown DH (1983) Factors affecting urease activity in the lichen *Evernia prunastri* (L). Ann Bot 52:261-264

Luque I, Flores E, Herrero A (1994) Molecular mechanism for the operation of nitrogen control in cyanobacteria. EMBO J 13:2862-2869

Mackerras AH, Smith GD (1986) Urease activity of the cyanobacterium *Anabaena cylindrica.* J Gen Microbiol 132:2749-2752

Mahapatra IC, Mudholkar NJ, Patnaik H (1971) Effect of N, P and K fertilizers on the prevalence of algae under field conditions. Ind J Agron 16:19-22

Martinek RG (1969) Review of methods for determining urea nitrogen in biologic fluids. J Am Med Technol 31:678-696

McCarthy JJ (1972) The uptake of urea by marine phytoplankton. J Phycol 8:216-222

McLean RJL, Cheng K-J, Gould WD, Costerton JW (1985) Cytochemical localization of urease in a rumen *Staphylococcus* sp. by electron microscopy. Appl Environ Microbiol 49:253-255

Mobley HLT, Hausinger RP (1989) Microbial ureases: Significance, regulation, and molecular characterization. Microbiol Rev 53:85-108

Mulvaney RL, Bremner JM (1981) Control of urea transformations in soils. In: Paul EA, Ladd JN (eds) Soil biochemistry, vol 5. Marcel Dekker Inc, New York, pp 153-196

Okuda A, Yamaguchi M (1952) Algae and atmospheric nitrogen fixation in paddy fields. II. Relation between the growth of blue-green algae and physical or chemical properties of soils and effect of soil treatments and inoculation on the nitrogen fixation. Mem Res Inst Food Sci 4:1-11

Oliveira L, Antia NJ (1984) Evidence of nickel ion requirement for autotrophic growth of a marine diatom with urea serving as nitrogen source. Br Phycol J 19:125-134

Olson DL, Christensen GM (1982) Effect of selected environmental pollutants and other chemicals on the activity of urease (*in vitro*). Bull Environ Contam Toxicol 28:439-445

Pateman JA, Dunn E, MacKay EM (1982) Urea and thiourea transport in *Aspergillus nidulans*. Biochem Genet 20:777-790

Polacco JC, Holland MA (1993) Roles of urease in plant cells. Internatl Rev Cytol 145:65-103

Qin Y, Cabral MS (1994) Kinetic studies of the urease-catalyzed hydrolysis of urea in a buffer-free system. Appl Biochem Biotechnol 49:217-240

Rai AK (1989) Purification and properties of urease from a cyanobacterium *Anabaena doliolum*. FEMS Microbiol Lett 61:319-322

Rai AK, Rai UN (1977) Simultaneous utilization of nitrate and ammonium-nitrogen by *Scenedesmus obliquus* (Turpin) Kuetzing. Microbios Lett 5:13-18

Rai AK, Singh S (1987a) Kinetics and regulation of urea uptake in *Anabaena doliolum* and *Anacystis nidulans*. J Gen Appl Microbiol 33:471-479

Rai AK, Singh S (1987b) Urease of blue-green algae (cyanobacteria) *Anabaena doliolum* and *Anacystis nidulans*. Curr Microbiol 16:113-117

Rawson DM (1985) The effects of exogenous amino acids on growth and nitrogenase activity in the cyanobacterium *Anabaena cylindrica* PCC 7122. J Gen Microbiol 131:2549-2554

Rees TAV, Syrett PJ (1979) The uptake of urea by the diatom *Phaeodactylum*. New Phytol 82:169-178

Rinaudo G (1974) Fixation biologique de l'azote dans trois types de sols de riziére de Côte d'Ivoire. Rev Ecol Biol Sol 11: 149-168

Roger PA, Reynaud PA (1979) Ecology of blue-green algae in paddy fields. In: Nitrogen and rice. Internatl Rice Res Inst, Los Banos, Philippines, pp 289-309

Roon RJ, Levenberg B (1968) An adenosine triphosphate dependent avidin sensitive enzymatic cleavage of urea in yeast and green algae. J Biol Chem 243:5213-5215

Roon RJ, Levenberg B (1972) Urea amidolyase. 1. Properties of the enzyme from *Candida utilis*. J Biol Chem 247:4107-4113

Rosenstein I, Hamilton-Miller JMT, Brumfitt W (1980) The effect of acetohydroxamic acids on the induction of bacterial ureases. Invest Urol 18:112-114

Shapiro IA (1977) Study of urease activity in lichen species. Sov Plant Physiol 24:914-918

Simon RD (1971) Cyanophycin granules from the blue-green alga *Anabaena cylindrica*: A reserve material consisting of copolymers of aspartic acid and arginine. Proc Natl Acad Sci USA 68:265-267

Singh RN (ed) (1961) Role of blue-green algae in nitrogen economy of Indian agriculture. Indian Council Agricultural Research, New Delhi

Singh S, Rai AK (1990) Nickel-dependent growth and urea uptake in the cyanobacterium *Anabaena doliolum* and *Anacystis nidulans*. Ind J Exptl Biol 28:80-82

Singh S, Rai AK (1989a) Methionine sulphoximine induced changes in cyanobacteria, *Anabaena doliolum* and *Anacystis nidulans*. Ind J Exptl Biol 27:227-229

Singh S. Rai AK (1989b) Localization of urease in *Anabaena doliolum*. Curr Sci 58:859

Sumner JB (1926) The isolation and crystallization of the enzyme urease. J Biol Chem 69:435-441

Syrett PJ, Leftley JW (1976) Nitrate and urea assimilation by algae. In: Sunderland N (ed) Perspective in experimental biology. Pergmon Press, Oxford, pp 221-234

Than-Tun (1969) Effect of fertilizers on the blue-green algae of the soils of the paddy fields of Mandalay agricultural station, Union Burma. J Life Sci 2:257-258

Thompson JF, Muenster AME (1971) Separation of the *Chlorella* ATP-urea amidolyase into two components. Biochem Biophys Res Commun 43:1049-1055

Tilman D (1976) Ecological competition between algae. Experimental confirmation of resource-based competition theory. Science 192:463-465

Tilman D (1977) Resource competition between planktonic algae: An experimental and theoretical approach. Ecology 58:338-348

Turpin DH, Harrison PJ (1979) Limiting nutrient patchiness and its role in phytoplankton ecology. J Exp Mar Biol Ecol 39:151-156

van Baalen C, O'Donnell R (1978) Isolation of a nickel-dependent blue-green algae. J gen Microbiol 105:351-353

Vicente C, Palasi M, Estvez MP (1978) Urease regulation mechanisms in *Lobaria pulmonaria* (L) Hoffm. Rev Bryol Lichenol 44:83-89

Vicente C, Xavier Filho L (1979) Urease regulation in *Cladonia verticillaris* (Raddi) Fr. Phyton (Buenos Aires) 37:137-144

Ward AK, Wetzel RG (1980) interaction of light and nitrogen source among planktonic blue-green algae. Arch Hydrobiol 90:1-25

Watson CJ, Miller H, Poland P, Kilpatrick DJ, Allen MDB, Garrett MK, Christianson CB (1994a) Soil properties and the ability of the urease inhibitor *N*-(n-butyl) thiophosphoric triamide (nBTPT) to reduce ammonia volatilization from surface-applied urea. Soil Biol Biochem 26:1165-1171

Watson CJ, Poland P, Miller H, Allen MBD, Garrett MK, Christianson CB (1994b) Agronomic assessment and ^{15}N recovery of urea amended with the urease inhibitor nBTPT (N-(n-butyl) thiophosphoric triamide) for temperate grassland. Plant Soil 161:167-177

Watson CJ, Stevens RJ, Laughlin RJ (1990) Effectiveness of the urease inhibitor NBTPT (N-(n-butly) thiophosphoric triamide for improving the efficiency of urea for ryegrass production. Fertil Res 24:11-15

Weathers PJ, Chee HL, Allen MM (1978) Arginine catabolism in *Aphanocapsa* 6308. Arch Microbiol 118:1-6

Wetzel RG (ed) (1975) Limnology. WB Saunders Co, Philadelphia

Williams SK, Hodson RC (1977) Transport of urea at low concentrations in *Chlamydomonas reinhardi*. J Bacteriol 130: 266-273

Wöhler F (1828) Ann Phys (Leipzig) [2] 12: 253, through Lipman TO (1964) J Chem Ed 41: 452

Wootton IDP (ed) (1974) Microanalysis in medical biochemistry. Churchill Livingstone Edinburg London

Xavier Filho L, Vicente C (1978) Exo- and endourease from *Parmelia roystonea* and their regulation by lichen acids. Bolet Soc Brot Ser 2, 52: 51-65

Nitrogen Metabolism of the Symbiotic Systems of Cyanobacteria

JACEK PLAZINSKI

Fellow, Plant Cell Biology, Research School of Biological Sciences, Australian National University, GPO Box 475, Canberra City, ACT 2601, Australia

1 Introduction

A definition of symbiosis, the dynamic process which is the result of interaction between two partners, is by no means easy. The most fundamental feature of symbiotic interactions is the bidirectional exchange of metabolites and nutrients between host and symbiont. The symbiotic microorganisms which live inside the cells or tissues of plants are called endophytes, as opposed to endozoic organisms living inside cells or tissues of animals.

In addition to being widely distributed as free-living prokaryotes, certain filamentous nitrogen-fixing cyanobacteria are able to form symbiotic associations with eukaryotic organisms. In the (endo)symbiotic state, many cyanobacteria are able to locate and invade the tissue of their hosts. The number of genera of the major plant groups which are able to develop such associations is strictly limited. However, the range of plants with which cyanobacteria can form a symbiosis is very wide. It encompasses algae, fungi, mosses, liverworts, ferns, cycads (representing a class of gymnosperm), and angiosperms (STEWART et al. 1983). Symbioses involving the cyanobacteria (cyanobioses) appear to be the most simple, though even that with fungi (forming lichens) must involve a series of complex interactions. In general, a very few species form specific associations with various plants, and in such associations the cyanobacteria (cyanobionts) become modified physiologically, morphologically, and biochemically. The location and morphology of the structures in which the algal symbionts are to be found range from root glands (nodules) to leaf cavities. The cyanobacterial symbiont is usually a species of *Nostoc, Anabaena, Calothrix, Fischerella* and *Scytonema* all of which are heterocystous forms (STEWART et al. 1983). Usually, the proportion of heterocysts (specialised cells where nitrogenase is synthesised) to vegetative cells is higher in the symbiotic form than in the free-living cyanobacteria and this is determined by the nitrogen status. Nitrogen is fixed by the heterocysts and diffuses into the vegetative cells. Compared to free-living forms, the cyanobacteria in association with a photosynthetic eukaryotic partner have a lower growth rate and a diminished capacity to assimilate ammonium and photosynthetically reduced CO_2.

Among the cyanobacterial symbioses there is a great diversity of interactions, ranging from the lichens when the cyanobiont photosynthesizes and provides the host with both carbon and nitrogen, through to the association with *Azolla* where photosynthesis takes place but carbon is supplemented by the host, to the situation in cycads and in *Gunnera* where the endophyte is probably totally dependent upon the host for carbon (SILVESTER 1976).

In the mutualistic cyanobiosis, fixed nitrogen is given up to the host and it is expressed by the high proportion of heterocysts in all these associations. Unfortunately, nitrogen metabolism processes in the cyanobacterial-plant symbioses are poorly understood compared to other systems such as the *Rhizobium*-legume symbiosis. There is a significant gap in the knowledge of nitrogen metabolism in cyanobioses, yet some associations, when properly understood and controlled, could be of great benefit to world agriculture. This chapter highlights the principle features of the nitrogen metabolism of a diverse range of cyanobacterial-plant symbiotic systems.

2 Symbiotic cyanobacteria - cyanobionts

The nitrogen-fixing cyanobacteria, the members of Cyanophyceae which develop into endophytic symbioses have unicellular or filamentous forms (Table 1).

Cyanobacteria involved in symbiotic associations are usually capable of growing and fixing nitrogen independently of their plant host. In general, the cyanobacterial partner capable of fixing nitrogen in the free-living state will maintain or even increase this function while in the symbiotic state. *Nostoc* strains (RIPPKA et al. 1979), are involved in most of the associations, but a number of other cyanobacterial genera, including unicellular ones, have been reported (RAI 1990). The presence of *Calothrix* in a few species of cycads has been reported (GROBBELAAR et al. 1987). In the *Azolla* association the main cyanobiont has been classified as *Anabaena azollae* (STRASBURGER 1873), as *Nostoc* sp. (MEEKS et al. 1988, PLAZINSKI et al. 1990a), and *Trichormus* (BERGMAN et al. 1992). In the *Gunnera* association the cyanobiont has been assigned to the genus *Nostoc* (LINDBLAD et al. 1989).

Most cyanobacteria which are able to form symbiotic associations have motile stages (hormogonia), the feature that may initiate contact between the eventual partners. The motile stages of cyanobionts were observed in the cavities of *Anthoceros*, in the glands of *Gunnera*, in the *Nostoc* colonies of *Blasia*, and the *Anabaena* colonies on the young leaves of *Azolla* ferns (reviewed by STEWART 1980).

In all cases of symbiosis the cyanobiont enters some permanent, normally internal, endophytic state during which it lives in a microhabitat (an intercellular or extracellular space or pocket) protected against external environmental factors by the host tissues.

In extracellular symbioses the endophytes are situated in preformed cavities surrounded by the cell walls of host cells. In intracellular symbioses, the penetrating symbionts may initially be surrounded by a cell wall deposit as in infection threads of penetrating *Rhizobium* during the initiation of root nodules in legume plants.

All endophytic symbionts are surrounded by a cytoplasmic membrane of predominantly host origin. The function of these perisymbiont membranes is illustrated by the universal presence of an active H^+-ATPase (SMITH and SMITH 1990), which indicates the presence of transport systems. In contrast, the membranes around endophytic parasites have no active H^+-ATPase (QUISPEL 1992).

The cyanobioses are relatively specific, and among the changes which occur in the endophytes are increases in the size of vegetative cells, changes in the ultrastructure of the vegetative cells, reduction in the length of filaments, an increased number of heterocysts (STEWART et al. 1983), reduced activities of glutamine synthetase (GS) and glutamate synthase (GOGAT), and reduced capacity to assimilate ammonia (STEWART and RODGERS 1977).

In the cyanobacterial vegetative cells much of the protoplast is occupied by the components of the photosynthetic apparatus, which is located in a series of membranous sacs or thylakoids (FOGG et al. 1973). In general, all cyanobacteria synthesise glycogen (FOGG et al. 1973), and cyanophycin (SIMON 1971) as reserve materials.

Heterocysts, the exclusive sites of nitrogen fixation, can readily be distinguished from the vegetative cells by three structural properties: 1) weak pigmentation; 2) thicker outer envelope; and 3) the presence of a polar granule near each junction with a vegetative cell (STANIER and COHEN-BAZIRE 1977). Heterocyst differentiation is subject to inductive control. A mature heterocyst is a terminal state as it has lost the ability to de-differentiate (WALSBY and NICHOLS 1969, WILCOX 1970, BUIKEMA and HASELKORN 1991).

In general, the cyanobacterium's symbiotic existence is reflected more in quantitative changes rather than qualitative. The functioning of the symbiosis is demonstrated by an effective fixation of dinitrogen. In the symbiotic state the fixation rate per cyanobacterial cell is quite rapid, although other metabolic activities such as growth and cell division proceed rather slowly, and a high proportion of the nitrogen fixed by the cyanobacteria is released from the cells, directly benefiting the host.

3 Symbiotic systems with cyanobacteria - cyanobioses

In most cases, the development of a cyanobiosis involves morphological changes in both the host and the cyanobiont. The overall effect of these changes is that the cyanobiont becomes unable to provide its own requirements for carbon and lives heterotrophically on carbohydrate provided by the host or a photosynthetic partner. Cyanobacteria normally colonize structures which already exist in their host plants, but occupation of these areas frequently results in morphological changes of the host. For example, in the bryophytes the cyanobacterium occupies cavities in the gametophyte thallus. In the vegetative sporophyte of *Azolla* fern the cyanobacterium also inhabits cavities but these are located in the chlorophyllous dorsal leaves (BERGMAN et al. 1992). In the *Azolla* cavities, besides cyanobacteria, a number of different bacteria have been found (WALLACE and GATES 1986, PETRO and GATES 1987, NIERZWICKI-BAUER and AULFINGER 1990, PLAZINSKI et al. 1990b). These bacteria possess nitrogenase components, and are therefore potentially diazotrophic (LINDBLAD et al. 1991). It has been suggested that these bacteria are the third partner of the *Azolla-Anabaena* symbiosis (CARRAPICO 1991). In the higher plant symbioses cyanobacteria occupy only non-chlorophyllous tissues. For instance, in cycads, specialized coralloid roots are occupied, whereas in angiosperms (*Gunnera* spp.), cyanobacteria colonize specific glands (nodules?) developed on the stem (BONNETT and SILVESTER 1981, SCHMIDT 1991). The *Gunnera-Nostoc* association is the only high plant symbiosis where cyanobacteria are found intercellularly (Table 1).

Table 1. General characteristics of the endophytic cyanobacterial symbioses

Symbiotic partners				N_2 fixation of symbiont	Form of association	Specificity of symbiont	Isolation of cyanobiont in culture	Reconstitution of symbiosis
Host		Endophyte						
Plant group	Genus	Major	Minor					
Algae	*Rhizosolenium*	*Calothrix* sp. *Richelia* sp.	UN	Yes	intercellular	Yes	Yes	ND
Fungi	Lichens	*Nostoc* sp.	UN	Yes	extracellular	Yes	Yes	ND
		Scytonema sp.	UN	Yes	extracellular	Yes	Yes	ND
		Fischerella sp.	UN	Yes	extracellular	Yes	Yes	ND
		Calothrix sp.	UN	Yes	extracellular	Yes	Yes	ND
Bryophytes	*Anthoceros* sp.	*Nostoc* sp.	cyanobact.[1]	Yes	extracellular	No	Yes	Yes
	Blasia sp.	*Nostoc* sp.	UN	Yes	extracellular	No	Yes	ND
	Sphagnum sp.	*Nostoc* sp.	UN	Yes	extracellular	No	Yes	ND
Pteridophytes	*Azolla* sp.	*Anabaena* / *Nostoc azollae* *Trichormus* sp.	 *Nostoc* sp.[1] *Arthrobacter*[2] Eubacteria[2]	Yes	extracellular	No	Yes[3]	Yes[4]
Gymnosperms	*Cycas* *Macrozamia* *Bowenia* *Ceratozamia*	*Anabaena* or *Nostoc* sp. *Calothrix* sp.	UN	Yes	extracellular	No	Yes	Yes
Angiosperm	*Gunnera* sp.	*Nostoc punctiforme*	UN	Yes	intercellular	Yes	Yes	Yes

UN - unknown

ND - not done, i.e., lack of published records of successful reconstitution of symbiotic association.

1- Recent studies showed that most of the isolates from endophytic symbioses represent a minor cyanobiont (NEWTON and HERMAN 1979, FRANCHE and COHEN-BAZIRE 1985, NIERZWICKI-BAUER and HASELKORN 1986, LIN et al. 1988, MEEKS et al. 1988, LIU et al. 1989, ZIMMERMAN et al. 1989, PLAZINSKI et al. 1990a, GEBHARDT and NIERZWICKI-BAUER 1991).

2- Several studies showed the presence of bacteria in the cavities of *Azolla* (NEWTON and HERMAN 1979, GATES et al. 1980, WALLACE and GATES 1986, PETRO and GATES 1987, CAIOLA et al. 1988, NIERZWICKI-BAUER and AULFINGER 1990, PLAZINSKI et al. 1990b, CARRAPICO 1991). There have been suggestions that these microorganisms are the third partner of the symbiosis, however as yet their role/function in the association has not been shown (NIERZWICKI-BAUER and AULFINGER 1990, PLAZINSKI et al. 1990b, CARRAPICO 1991).

Yes[3]- Only a minor cyanobiont has been successfully isolated (NEWTON and HERMAN 1979, NIERZWICKI-BAUER and HASELKORN 1986, LIN et al. 1988, LIU et at 1989, GEBHARDT and NIERZWICKI-BAUER 1991).

Yes[4]- Reconstitution of the *Azolla-Anabaena* symbiosis has been achieved by an unconventional method of grafting an indusium (a cap which contained *Anabaena azollae* filaments on its inner surface) onto the original position of the decapitated (indusium-free) megaspore of the same or another species of fern prior to germination (LIN et al. 1988, LIU et at 1989, PLAZINSKI et al. 1989).

Soon after the cyanobacterium has penetrated the host, a number of factors come into play, most of which are poorly understood. There is probably some specificity exhibited between the host plant and its cyanobiont (e.g., some, but not all, species of *Anabaena/Nostoc* can associate with *Azolla).*

When the host plant is capable of photosynthesis or when photosynthetic organisms coexist with the cyanobacteria inside the plant (as in some lichens) the frequency of heterocysts is increased at the expense of vegetative cells. This process of cell differentiation results in an increased capacity to fix nitrogen. Ammonia, the product of nitrogen fixation is assimilated by means of glutamine synthetase (GS) and glutamate synthase (GOGAT). In some cyanobioses (e.g., bryophytes and ferns), the activity of GS enzyme is lower in symbiotic than in free-living cyanobacteria. Therefore, the majority of the total fixed nitrogen is exported to the host plant, whereas only a small fraction of the fixed nitrogen is assimilated by the cyanobacterium. Such inbalanced nitrogen status, frequently found in the active symbiosis, makes the cyanobiont deficient in nitrogen that in turn initiates the development of more heterocysts (WALSBY and NICHOLS 1969, STEWART et al. 1983).

3.1 Miscellaneous symbioses

There are a number of associations known between cyanobacteria and microorganisms which lack chlorophyll. In these associations, called endocyanoses, the cyanobacterium is always intercellular (MILLBANK 1974, STEWART et al. 1983).

Geosiphon pyriforme is an association between *Nostoc sphaericum,* a cyanobacterium which is known to fix nitrogen in pure culture, and a phycomycete fungus, which becomes vesicular in the symbiotic state (PANKOW and MARTENS 1964). This symbiosis is autotrophic where the cyanobacterium functions as a chromatophore and is also able to fix nitrogen.

A heterocystous, unicellular cyanobacteria *Richelia* and *Calothrix* (Table 1) form an association with the marine planktonic alga *Rhizosolenium* (STEWART et al. 1980). Detected nitrogenase activity is light dependent and can be inhibited by ammonia.

3.2 Associations with fungi - Lichens

Lichens are regarded as highly selective morphological forms of some 18,000 species (GALLON and CHAPLIN 1987), from which about 8% contain a nitrogen-fixing cyanobacterium as the primary or secondary photobiont (STEWART et al. 1983, MILLBANK 1984).

In general, lichens are symbioses of fungi (ascomycetes and basidiomycetes) with green algae as the symbiont. Potential partners for an initiation of the lichen symbiosis are fungal spores as well as green algal or cyanobacterial cells. The most common cyanobacteria found in lichens are species of *Nostoc, Scytonema, Fischerella, Calothrix* and *Gloeothece* (HITCH and MILLBANK 1975, STEWART et al. 1983). The vegetative body of a lichen is known as the thallus and is composed of two organisms, a fungal component (mycobiont) and a green alga or cyanobacterium (photobiont). Lichens are long-lived organisms and have characteristics which are quite distinct from those of either partner.

In most symbioses (the two-membered association), the cyanobacterial filaments are either confined to a single layer or more randomly dispersed throughout the fungal hyphae. There are also a few species, such as *Peltigera aphthosa* and *Stereocaulon vesuvianum,* that have a triple symbiosis (STEWART et al. 1983). In the three-membered association the main part of the thallus is composed of an association of the fungus with a green alga and special structures, cephalodia, contain cyanobacteria which fix nitrogen for the thallus as a whole.

The structure of the lichen thallus varies from *Collema,* where the fungal hyphae and cyanobacterial filaments intermingle through the thallus, to *Peltigera canina,* where the cyanobacteria are confined to a thin layer beneath the upper cortex of fungal hyphae (ISAAC 1992). In the lichens the form is determined by both the fungus and the alga, so it is not possible to ascertain whether any particular change in form is concerned with the cyanobacterial association or with nitrogen fixation.

Lichens that contain cyanobacteria are of ecological rather than of economic importance. The possession of nitrogen-fixing ability by some lichens must be an advantage; however the cyanobacterial associations with fungi do not seem to have any great advantage over the green algal associations except in conditions of extreme nitrogen deficiency. In general, the nitrogen input to the habitat in which they live is rather small.

The fungus depends upon the cyanobacterium for both carbon and nitrogen when the sole partner is a cyanobacterium, but for nitrogen alone when the cyanobacteria are confined to cephalodia. The carbon which is fixed by the photosynthetic activities of the photobiont passes to the fungus and fully sustains the needs of the mycobiont partner. There is, however, very little evidence for any exchange of metabolites between the fungus and the photobiont. It has therefore been suggested that the relationship between the two partners is a controlled parasitism of the photobiont by the fungus, mediated only by the rapid proliferation of the algal cells (AHMADJIAN 1982).

Lichens with cyanobacterial symbionts have the capacity to fix nitrogen (GALUN and BUBRICK 1984). In those lichen species which have cyanobacteria as the sole symbionts the amounts and rates of nitrogen fixation are quite large on a cellular basis and sometimes greater than those in free-living cyanobacteria of the same species. When the cyanobacteria are present in the main thallus the proportion of heterocysts is the same as in the free-living cyanobacteria. This is because the growth of the lichen is slow and nitrogen is not a growth-limiting factor. When the cyanobacteria are confined to the cephalodia, the nitrogen is exported to the main thallus (ISAAC 1992). This forms a large sink for the nitrogen, as it is required both for the fungus and its cyanobacterial partner. As the cephalodia form a small proportion of the whole organism, the nitrogen is depleted and as a consequence the proportion of heterocysts is high (HITCH and MILLBANK 1975).

In order that the fungus may benefit from the nitrogen-fixing ability of the cyanobacteria, the nitrogen metabolism of the latter becomes modified (STEWART et al. 1980, 1983). The enzyme concerned with the incorporation of nitrogen in free-living cyanobacteria is glutamine synthetase. The amount of this enzyme in the endophytic form of the cyanobacterium is considerably reduced (BERGMAN and RAI 1989) so that it is unable to incorporate much of the nitrogen that it has fixed. This is in contrast to the situation in the cyanobionts of other plants where the GS level in heterocysts is at least two-fold higher than in vegetative cells (LINDBLAD and BERGMAN 1986). Most of the fixed nitrogen is released as ammonia, which is subsequently converted and stored as glutamate in the fungus (reviewed by ISAAC 1992). Amounts of other nitrogen-metabolizing enzymes concerned with the further processing of nitrogen compounds for amino-acid synthesis are also reduced in the endophyte. The mechanism by which this reduction in nitrogen-metabolizing enzymes is achieved is as yet poorly understood.

3.3 Associations with bryophytes

It is now well established that nitrogen-fixing cyanobacteria can form associations with a small number of bryophytes (RODGERS and STEWART 1977, STEWART and RODGERS 1977). This the only group among the cyanobioses which occurs in temperate climates. Of these associations, the liverworts *Anthoceros, Blasia,* and *Cavicularia,* and the moss *Sphagnum* have been most extensively studied. In all of these systems it is relatively easy to separate the host from its endophyte and to grow both partners independently, thereby facilitating laboratory investigations into the symbiotic relationship.

In the liverworts, the cyanobionts are members of the genus *Nostoc.* The cyanobacterial cells develop on mucilage-filled cavities on the thallus undersurface. Once the cyanobiont is established it stimulates the formation of papillae within the cavity and the *Nostoc* filaments then ramify amongst these papillae, forming compact colonies. Most likely this arrangement facilitates the interchange of material between the endophyte and its host. The ability to stimulate the formation of papillae is a property only of the limited number of strains of *Nostoc* which are able to form successful symbiotic associations. Penetration can be effected by other cyanobacteria, or even by some eukaryotic algae, but no further development takes place in these cases.

During the establishment of the symbiosis, several morphological changes occur in the cyanobiont. The vegetative cells become larger and more rounded and the heterocyst frequency increases to as much as 43% (STEWART et al. 1983). In conjunction with these morphological changes, several biochemical modifications occur. The photosynthetic capacity of the cyanobiont (as measured by CO_2 fixation, O_2 evolution, and the level of accessory pigments for photosystem II) becomes lower. Thus, the cyanobacterium may become dependent upon a supply of carbohydrate from its host. Ammonia assimilation by both host and cyanobiont occurs by means of the GS/GOGAT pathway (MEEKS et al. 1985a). Compared to the free-living form, the endophytic cyanobacterium shows a decreased capacity to assimilate ammonia, as much as 90% of the nitrogen fixed being exported to the host in the form of ammonia (MEEKS et al. 1985a). Excretion of ammonia by the cyanobiont appears to result from a positive modification of the *Nostoc* to a symbiotic mode of existence.

Sphagnum mosses contain normal vegetative cells interspersed with dead, water-filled hyaline cells, and it is inside the latter that *Nostoc* is sometimes found. Mosses are frequently found in rather acidic habitats not normally favoured by cyanobacteria. However, a process of cation exchange between the moss and its environment probably provides a more alkaline situation inside the hyaline cells than exists outside them (GRANHALL and HOFSTEN 1976).

Nitrogen fixation by the association is confined to the outer areas of the moss where the light intensity is high enough to permit the *Nostoc* to grow phototrophically. The rate of nitrogen fixation is both temperature- and light-dependent, but relatively insensitive to pH over the range pH 4-7 (BASILIER and GRANHALL 1978). Estimates of the annual rate of nitrogen fixation suggest that about 94 kg N/ha/yr might be fixed (GALLON and CHAPLIN 1987). Thus mosses may make a considerable contribution to the nitrogen economy of those environments where they occur.

3.4 Associations with the Pteridophyte - *Azolla*

Azolla is a genus of heterosporous aquatic ferns that grow on the surface of fresh water ponds, lakes, streams or irrigation channels. It has a worldwide distribution but occurs most abundantly in the tropics. Individual species have been distributed by man by natural means to temperate as well as tropical and subtropical areas (LUMPKIN and PLUCKNETT 1982). *Azolla* is usually found associated with the heterocystous, nitrogen fixing cyanobacterium *Anabaena azollae*, which occupies mucilage-filled cavities on the ventral surface of the dorsal lobe of each leaf (PETERS et al. 1980). The *Anabaena* infects the fronds of the fern at an early stage of the plant's development and becomes enclosed in a pocket within the frond (PETERS 1977), Unlike the legume-*Rhizobium* symbiosis, the *Azolla* host and the *Anabaena* symbiont normally remain together through successive cycles of vegetative and sexual reproduction. The symbiosis can confer high rates of nitrogen fixation and biomass production, hence *Azolla-Anabaena* is an effective green manure for flooded crops and has been used as biofertilizer in rice-growing for centuries in China and Vietnam (MOORE 1969, LUMPKIN and PLUCKNETT 1982). Recently *Azolla* has been successfully used as biofertilizer for green garden crops (MILICIA and FAVILLI 1992). The growth of *Azolla* appears to be temperature sensitive although the degree of sensitivity may vary from species to species (WATANABE and ROGER 1984). The use made of *Azolla* in rice agriculture may therefore depend upon the climatic conditions prevailing in a particular location.

Efforts are being made to improve the desirable agronomic attributes of this symbiotic association by creating host-symbiont combinations other than those available in natural populations (LIN and WATANABE 1988, LIN et al. 1988, PLAZINSKI et al. 1989). Under natural conditions the same symbiont is transmitted to successive generations by filaments which are carried in the megasporocarp: there is, in fact, maternal inheritance of the *Anabaena*. Production of recombinant partnership involves breaking this strict association, re-introducing a new cyanobacterial strain and identification of new partners (LIU et al. 1989, PLAZINSKI et al. 1989, ZIMMERMAN et al. 1989). This system is one of the most studied endophytic symbioses of cyanobacteria.

The relationship between the fern and its associated cyanobacterium is assumed to be one in which photosynthate is actively transferred from the host to the endophyte, whilst fixed nitrogen is translocated from the endophyte to the fern. The availability of photosynthate provided by *Azolla* may decrease the dependence of the *Anabaena* on its own photosynthate and thus permits the increase in heterocysts frequency which occurs in *A. azollae* within the leaf cavities. In this respect, heterocyst frequency as high as 40% have been recorded (reviewed by PLAZINSKI 1990). There is some evidence that increased heterocyst frequency is not dependent upon a supply of photosynthate from a symbiotic partner. It is therefore possible that at least some of the benefits obtained by the cyanobacterium in forming the symbiotic association relate to the physical environment within the leaf cavities rather than to any provision of soluble materials from the host.

It has been shown that *Anabaena azollae*, immediately after isolation from *Azolla*, fixes nitrogen at a rate at least equal to free-living cyanobacteria (RAY et al. 1978, PETERS et al. 1980, KIKUCHI et al. 1984). There is much evidence that the symbiotic *Anabaena* releases a substantial portion of its newly fixed nitrogen as ammonium (MEEKS et al. 1985b). Once assimilated, fixed nitrogen is transported from mature leaf cavities to the growing sections of the fern mainly in the form of glutamine, glutamate, ammonia and an unidentified nitrogen-containing compound (PETERS et al. 1985). The nitrogen assimilation is conducted by the GS/GOGAT pathway (MEEKS et al. 1985b) with little or no synthesis of glutamate by glutamate dehydrogenase (GDH). The little biosynthetic activity of GDH in the isolated *Anabaena* indicates that all of GDH activity is associated with the hair cells of the cavity (PETERS et al. 1985). The nitrogen fixation and NH_4^+ assimilation are not tightly coupled as they are in free-living cyanobacteria (STEWART and ROWELL 1975, WOLK et al. 1976), and GS is repressed in symbiotic *Anabaena* (STEWART et al. 1980, ORR and HASELKORN 1982).

The close relationship between photosynthesis and nitrogen fixation in *Azolla-Anabaena* symbiosis is well documented (PETERS et al. 1980, reviewed by PETERS and MEEKS 1989). Photosynthesis is the ultimate source of ATP, reductant, and carbon skeletons used in nitrogen fixation. A number of studies strongly indicate that reductant from endogenous reserves of photosynthate and photophosphorylation are the primary driving forces for nitrogenase activity in the light; and that dark, respiratory-driven activities may be ATP limited (PETERS et al. 1980, reviewed by PETERS and MEEKS 1989, PLAZINSKI 1990). Interactions between photosynthesis and nitrogenase activity have been demonstrated by comparing action spectra for nitrogenase catalyzed C_2H_2 reduction in the *Azolla caroliniana-Anabaena* association and isolated cyanobiont (TYAGI et al. 1981).

The contribution that the *Azolla/Anabaena* association may make to the nitrogen economy of its environment may be quite considerable, values in the region of 1-2 kg N/ha/day being reported (WATANABE 1977).

3.5 Associations with gymnosperms - cycads

Within the gymnosperms, the family Cycadaceae consists of about nine genera and some 150 species, all of which produce coralloid roots which are morphologically distinct from the normal root system. The coralloid roots are modified lateral roots which branch dichotomously near or even above the soil surface. In about one-third of the species, the coralloid roots contain nodule-like structures which are inhabited by heterocystous cyanobacteria, variously identified as *Nostoc* or *Anabaena*. This makes cycads unique among gymnosperms. The mode of entry of the cyanobacteria into the root nodules has been debated but is probably via a break in the dermal layer. In most cases, the endophyte is extracellular but a few instances of an intercellular location have been recorded.

Nitrogen fixation by cycads has been reviewed by STEWART (1980). Cyanobacteria isolated from the nodule are capable of freely fixing nitrogen though one such isolate, *Nostoc macrozamia* (obtained from the cycad *Macrozamia*), has since lost the ability to freely fix N_2. As seems to be case in most other cyanobacteria isolated from symbiotic associations, heterotrophic growth is possible. In the case of *Nostoc macrozamia*, this can be supported by either glucose or fructose. Within the nodule, photosynthesis by the endophyte may be depressed, partly due to the soil covering the roots and partly due to tannins which make the roots opaque. Thus, growth of the endophyte within the root is probably at the expense of carbohydrate provided by the host.

When nodule-containing coralloid roots of *Macrozamia* were incubated with $^{15}N_2$, rates of N_2 fixation comparable to that of excised nodules of soybean were observed. The fixed ^{15}N became distributed throughout the plant except for the leaves. It is therefore clear that the endophyte can provide its host with fixed nitrogen.

Although the cycads were once of worldwide distribution, they now occur naturally only in limited areas. There are substantial variations in the rate of N_2 fixation with season but it has been estimated that the total amount of N_2 fixed can be as high as 18 kg N/ha/yr.

3.6 Associations with the angiosperm - *Gunnera*

Gunnera comprises some 50 species in the Haloragaceae, the water milfoil family. It is distributed mainly in the tropical and subtropical regions. The species are rhubarb-like herbaceous plants, ranging from small individuals with leaf blades of less than 10 cm to large upright ones with leaves of 2 to 3 m. *Gunnera* is the only known angiosperm genus which is capable of developing a symbiotic, nitrogen-fixing association with cyanobacterium *Nostoc punctiforme* (BECKING 1976, SILVESTER and MCNAMARA 1976). It is one of the few symbioses where the cyanobionts are found mainly intercellulary (NEUMANN et al. 1970, SODERBACK et al. 1990). The *Nostoc* occupies mucilage-filled glands situated at the base of the leaves. These glands develop whether or not the cyanobacterium is present to occupy them. However, as is the case with the *Rhizobium*/legume association, the endophyte is separated from its host's cytoplasm by a membrane of host origin. Both host and endophyte are capable of independent existence, but *Nostoc* is capable of providing the total requirements of the host plant for fixed nitrogen (BECKING 1976, BONNETT and SILVESTER 1981, SODERBACK et al. 1990).

The *Nostoc* enters glands found on the stem in the axils of the leaves. There are three such glands associated with each petiole base: one situated centrally on the stem above the emergence of the petiole and the other two lying laterally on either side of the petiole

(SILVESTER and MCNAMARA 1976). The *Nostoc* grows into the stem tissue in a fan-like form so that cells of *Nostoc* invade host cells well within the stem tissue. During the infection process the filaments invade the host cells and became intracellular, although they are separated from the host cell's contents by the plasmalemma. After invading the host cells, a large part of subsequent growth involves formation of heterocysts. The development of heterocysts continues through the life of the plant cell until cell senescence, when about 60% of the *Nostoc* cells are heterocysts.

During the establishment of the *Gunnera/Nostoc* symbiosis, the cyanobacterium undergoes a number of differentiation processes (JOHANSSON and BERGMAN 1992). Firstly, *Nostoc* differentiates motile, non-heterocystous hormogonia (straight, short motile filaments), which subsequently redifferentiate into aseriate filaments with an increasing proportion of heterocysts. Secondly, it develops from a photoautotrophic organism fixing nitrogen for its own needs, into a heterotrophic supplier of combined nitrogen. These modifications must be imposed upon the cyanobiont by the host through some signals synthesised in the gland and present in the mucus. Phenolics are considered to play an important role as their presence has been detected in the glands of *Gunnera* (SCHMIDT 1991).

In the earlier stages, when heterocysts form 20-40% of the *Nostoc* cells, the high proportion of nitrogen fixing cells provides very active nitrogen fixation, and it has been estimated that this is several times the rate of nitrogen fixation of free-living *Nostoc* in which about 5% of the cells are heterocysts.

The changes which occur in *Nostoc punctiforme* when it becomes involved in the association had been reviewed by SILVESTER (1976). Using a variety of methods, he estimated that the specific activity of nitrogenase was some five to ten times greater in endophytic *Nostoc* than in the free-living cyanobacterium, this marked difference probably being due to the increased heterocyst frequency that was observed in the endophytic cyanobacterium. Thus, although the heterocyst frequency in free-living *N. punctiforme* is usually about 5%, in association with *Gunnera* it ranges from 23% in nodules near the apex of the plant to 65% in more mature nodules further from the apex. This increased heterocyst frequency, accompanying as it does a decrease in the proportion of vegetative cells, decreases the amount of carbohydrate that can be provided by photosynthesis, this amount being decreased still further by the location of the cyanobacterial colonies inside the gland where they may be substantially shielded from light (SODERBACK et al. 1990).

The glands that contain *Nostoc* are shaded by the petiole and by other leaf bases, and the endophyte is also covered by a number of host cells, so that there is very little light for photosynthesis. However, because phycocyanin, a pigment that is associated with photosynthesis II, is absent from both the heterocysts and the vegetative cells, *Nostoc,* in symbiosis, must live heterotrophically on carbon fixed by the host. When *Nostoc* is separated from the host it is found that light can stimulate nitrogen fixation. This shows that photosynthesis I is operative even though it cannot be used in the symbiotic conditions. The mechanism by which nitrogen is transferred to the host has not so far been investigated, but by analogy with the other symbioses that have been considered, one would expect it to involve the suppression of glutamine formation by the endophyte.

When *Nostoc* is freshly isolated from *Gunnera,* neither CO_2 fixation nor O_2 evolution occur to any great extent. However, this observation cannot be explained solely in terms of the low ratio of vegetative cells to heterocysts. Rather it is probably due to the lack of significant amounts of the accessory pigments for photosystem II in those vegetative cells that remain. Thus, in symbiotic association, *Nostoc* has a greatly decreased capacity for photosynthesis and probably grows heterotrophically at the expense of carbohydrate provided by the host. The exact nature of this carbohydrate has not yet been established by *N. punctiforme* has been shown to grow heterotrophically in free-living culture, using fructose as the sole source of carbon (see SILVESTER 1976). The contribution of this symbiosis to the nitrogen economy of the environment in which it occurs may be considerable as estimates range from 12 kg N/ha/yr (BECKING 1976) to 72 kg N/ha/yr (SILVESTER 1976).

4 Comparisons between endophytic and cultured cyanobacteria

Comparisons between the freshly isolated endophytes and their cultivated strains may give valuable information about the characteristic aspects of the endophytic forms and may provide some indications about the factors which are responsible for these differences. In general it is difficult to ascertain that the cultures of isolated endophytic colonies really represent the same organism which formed the endophytic cells. It must be realized that endophytic forms represent a highly adapted form of microorganism which is susceptible to sudden changes in the environment during isolation. The necessity for a critical consideration of the identity of the cultivated strains is best illustrated for symbioses with cyanobacteria. In most hosts, the *Nostoc* or *Anabaena* is easily isolated and reinfections are possible, even with originally nonsymbiotic strains, showing a high, though not absolute, promiscuity of the isolated strains for their hosts.

Cyanobionts from all endophytic symbioses except *Azolla* have been successfully isolated and cultured in the laboratory (MEEKS 1990, LINDBLAD and BERGMAN 1990, BERGMAN et al. 1992). Various *Nostoc* strains were found to reconstitute *Anthoceros* and *Blasia* symbioses.

In the majority of symbioses the host is infected *de novo* by the cyanobiont. One exception is *Azolla* where the cyanobiont never leaves the host because there is continuity of the symbiosis throughout the life cycle of the fern (PETERS and MEEKS 1989). Although many strains of *Anabaena azollae* have been isolated, reinfections of symbiont-free *Azolla* plantlets is impossible or, at best, ineffective (LIN et al. 1988). The enzymatic, antigenic, morphologic and molecular differences of the isolated and cultivated strains with isolated suspensions of the endophytic forms are so great that either the cultures were markedly changed during cultivation or belonged to a minor population, differing from the dominant endophytic form (LIN et al. 1988, MEEKS et al. 1988). The question one may ask is whether the isolated strains are not representative of the major endophytic population or are significantly changed during cultivation. At least in an *Azolla-Anabaena* symbiosis the former seems to be the case.

In many symbionts there are no obvious differences between the endophytic and cultivated forms. Endophytic cyanobacteria may have somewhat larger dimensions and do not form spores, even though they are abundant in cultures (GALUN and BUBRICK 1984). They may have modified cell wall constituents as is evident from reaction with lectins or antigenic cross-reactions in algae from lichens (BUBRICK et al. 1982). Generally in symbioses with cyanobacteria, the photosynthetic pigments are usually still present in the endophytic cells, albeit at reduced level, even when they live entirely heterotrophically (TREDICI et al. 1988).

5 Transfer of fixed nitrogen in symbiotic systems

Symbiotic nitrogen fixation by cyanobacteria occurs at localized sites within the host organism. Effective connections between the site of fixation and the body of the host are necessary to remove the products of nitrogen fixation rapidly. This is achieved in various ways in different symbiotic systems.

In *Azolla* fronds the cyanobacterium *Anabaena* occupies a mucilage-filled cavity lined with multicellular hairs (DUCKETT et al. 1975, CALVERT and PETERS 1981, GUNNING and PATE 1969). These have wall infoldings typical of transfer cells and thus appear to be special adaptations for the transfer of fixed nitrogen to the host plant, although similar structures are also present in cavities of uninfected fronds when *Azolla* is grown on combined nitrogen. The dense cytoplasm of the projecting cells is suggestive of intense metabolic activity. Presumably they are involved in the assimilation of ammonia excreted by *Anabaena* into the cavity.

A rather similar arrangement is found in liverworts such as *Blasia* or *Anthoceros,* where the bryophyte produces branched filaments which project into the *Nostoc*-containing mucilaginous cavity in the thallus. These filaments also have wall infoldings typical of transfer cells and the structure again presumably functions to permit rapid movement of solutes between the cavity and the liverwort thallus. In *Azolla* assimilation of ammonia by the fern occurs via the glutamine synthetase-GOGAT pathway. However, in lichens such as *Peltigera aphthosa*, assimilation of ammonia excreted by *Nostoc* is by fungal glutamate dehydrogenase. Alanine is synthesized subsequently from glutamate by the aminotransferase reaction and it is in this form that the fixed nitrogen is transferred to the rest of the thallus.

Cyanobacterial roots nodules (glands) are found among the cycads and in an association with the angiosperm *Gunnera* (see STEWART et al. 1980, 1983). In nodules such as those of *Encephalartos* or *Macrozamia* the cyanobacteria are located in mucilaginous spaces between the cells in the mid-cortex. The coralloid nodules are of indeterminate growth and are well supplied with vascular connections back to the transport system of the host. In the *Macrozamia-Abanaena* symbiosis, citrulline and glutamine are the main nitrogenous compounds translocated (HALLIDAY and PATE 1976, DIXON and WHEELER 1986). In general, nitrogen release is either due to a reduced GS synthesis as in lichens and *Azolla* or due to inactivation of GS as in *Anthoceros* (JOSEPH and MEEKS 1987). It appears that in all cyanobioses where the cyanobiont liberates the fixed nitrogen as ammonia, there is a repression of GS in heterocysts (BERGMAN et al. 1992).

6 Nitrogen fixation by cyanobacteria

Cyanobacteria possess both photosystems I and II and are therefore capable of photolysing water to provide electrons and protons. In fact, they are the only prokaryotes capable of oxygenic photosynthesis. All species so far examined lack a complete tricarboxylic acid (TCA) cycle due to the absence of an enzyme with 2-oxoglutarate oxidoreductase activity. However, most cyanobacteria are aerobes and possess a respiratory chain. They are therefore capable of respiring and synthesising ATP by oxidative phosphorylation.

Nitrogen fixation by heterocystous cyanobacteria has been mainly studied using free-living organisms from only two genera, *Anabaena* and *Nostoc*. Inevitably, results using these two organisms have been extrapolated to less well investigated symbiotic species.

Filaments of most heterocystous cyanobacteria consist of three types of cells: vegetative cells, heterocysts, and akinetes. In normal free-living, nitrogen-fixing cultures there are about 10 to 20 vegetative cells for each heterocyst. In biochemical terms, the vegetative cells resemble the chloroplasts of higher plants in that they possess both photosystems I and II and also the enzymes of the Calvin cycle and are therefore able to fix CO_2 using ATP and reductant generated by photosynthesis. However, the heterocysts lack photosystem II activity and cannot produce O_2 by photosynthesis, nor are they capable of CO fixation because they lack the key enzymes of the Calvin cycle (RAY et al. 1979).

Most workers now accept that the heterocyst is the site of aerobic N_2 fixation although there have been reports that, when grown under anaerobic conditions, the vegetative cells are also capable of fixing nitrogen (HASELKORN 1978, STEWART 1980).

The relationship between the heterocyst and the vegetative cells is one of mutual interchange of materials. The vegetative cell, lacking the ability to fix nitrogen, is dependent upon the heterocyst for a supply of carbohydrate. This relationship has been recently reviewed by BOTHE et al. (1984) and STEWART et al. (1985). The exact identity of the carbohydrate which is translocated to the heterocyst has not yet been established although maltose, sucrose, and erythrose have been suggested as likely candidates (see STEWART et al. 1985). Carbohydrate metabolism in the heterocyst has, in the past, been assumed to occur largely through the activity of the oxidative pentose phosphate pathway. In support of this view, the two key enzymes of this pathway (glucose-6-phosphate dehydrogenase and 6- phosphogluconate dehydrogenase) occur in the heterocyst at far higher activities than in the vegetative cell. However, more recent work has shown that the heterocysts of a free-living *A.cylindrica* also possess all of the enzymes of the glycolytic pathway (reviewed by HASELKORN 1978, GALLON and CHAPLIN 1987). Thus, pyruvate may be generated from monosaccharides by either pathway; the relative contribution of each is still unknown.

In most aerobic organisms, acetyl CoA (the product of pyruvate oxidation) is completely oxidised to CO_2 through the action of the tricarboxylic acid (TCA) cycle with the generation of reducing equivalents in the form of $NADH_2$. This reduced coenzyme is used to supply electrons for electron transport and oxidative phosphorylation. However, in cyanobacteria, the TCA cycle is incomplete owing to the lack of an enzyme capable of oxidising 2-oxoglutarate (BOTHE et al. 1984, STEWART et al. 1985). Thus, although limited amounts of $NADH_2$ can be generated by what remains of the cycle (particularly via isocitrate dehydrogenase), the main function of the cycle in heterocysts appears to be to generate carbon skeletons for ammonia assimilation. In this connection, because glutamate is required as the substrate for the glutamine synthetase (GS) reaction, it was believed until recently that glutamate was translocated from the vegetative cell. However, it has now been established that the heterocyst itself contains glutamate synthase (GOGAT) and is therefore able to synthesise (from 2-oxoglutarate and glutamine) the glutamate it requires for ammonia assimilation (see STEWART 1980). Nevertheless, the formation of 2-oxoglutarate requires the condensation of acetyl CoA with oxaloacetic acid to form citric acid and oxaloacetic acid is also required as a precursor of aspartic acid, a component of the storage polymer cyanophycin (a peptide consisting of alternating aspartic acid and arginine residues). In most aerobic organisms, oxaloacetate is regenerated from 2-oxoglutarate via the TCA cycle, but in cyanobacteria, because this is not possible, an alternative source of oxaloacetic acid is necessary (see GALLON and CHAPLIN 1987). The presence in the heterocyst of the enzymes phosphoenolpyruvate carboxylase (which carboxylates phosphoenolpyruvate to oxaloacetic acid) and an NADP-dependent malic enzyme (which carboxylates pyruvic acid to malic acid) provides a means whereby the production of oxaloacetic acid can be achieved. This observation, along with the finding that both GS and GOGAT occur within the heterocyst, support the current belief that the whole process of ammonia assimilation occurs entirely within the heterocyst and that glutamine is exported to the vegetative cell (HASELKORN 1978, STEWART 1980, BUIKEMA and HASELKORN 1991).

It is now generally believed that, in cyanobacterial heterocysts, ferredoxin is the natural electron carrier to nitrogenase, although under iron-deficient growth conditions flavodoxin is produced and this can substitute for ferredoxin in N_2 fixation (reviewed by GALLON and CHAPLIN 1987).

Heterocysts lack photosystem II activity and, therefore, the generation of reductant from water by non-cyclic electron transport in the heterocyst is not possible. However, in heterocysts ferredoxin might be reduced directly in the light by non-cyclic electron transport involving photosystem I, provided that a suitable source of electrons is available. H_2, through the action of an uptake hydrogenase, is a possible source of reductant for N_2 fixation by heterocystous cyanobacteria. Probably a reductant for N_2 fixation is generated by more than one of the mechanisms, but the relative contribution of each is as yet unknown (BOTHE et al. 1984).

Nitrogenase can be irreversibly inactivated by being exposed to O_2, so all N_2 fixing cyanobacteria must protect their nitrogenase not only from atmospheric O_2 but also from oxygen generated intracellularly by photosynthesis. An N_2-fixing cyanobacterium can protect its nitrogenase from atmospheric O_2 by one or more ways: having a physical barrier to O_2, metabolic removal of O_2 and H_2O_2 (e.g., respiration), and conformational protection or continuous synthesis of nitrogenase (BURRIS 1979, YOUNGMAN 1984).

Almost all heterocystous cyanobacteria fix N_2 aerobically in the light. Protection from photosynthetically derived O_2 can also be achieved by separating the processes of N_2 fixation and photosynthetic O_2 evolution, either in space (spatial separation) or in time (temporal separation). The heterocystous cyanobacteria, by confining photosynthetic O_2 evolution to the non-nitrogen fixing vegetative cells achieve a spatial separation of nitrogenase and photosynthetically-derived O_2. Cyanobacteria, therefore, minimise the problem of photosynthetically produced O_2 either by confining N_2 fixation to cells which do not evolve O_2 or by confining N_2 fixation to periods of minimal photosynthetic activity.

The intracellular concentration of nitrogenase can be controlled *in vivo* by repression of nitrogenase synthesis and by degradation of the pre-existing enzyme. In *Gloeothece* it might be glutamine itself which represses nitrogenase synthesis (THOMAS et al. 1982). This is a repression of nitrogenase by NH_4^+ and its derivatives. Nitrogenase synthesis can also be repressed by nitrate or nitrite, O_2 or other factors such as molybdenum, cyclic nucleotides and nitrogen starvation (HOM et al. 1980, GALLON and CHAPLIN 1987). Light might also be involved in the repression of nitrogenase (GALLON and CHAPLIN 1987).

In all cyanobacterium-higher plant cyanobioses the nitrogenase activity varies from younger to older symbiotic tissues (BRAUN-HOWLAND and NIERZWICKI 1990, LINDBLAD and BERGMAN 1990, SODERBACK et al. 1990), however, the increased nitrogenase activity is not linear with the increase in heterocyst frequency (SODERBACK et al. 1990, BERGMAN et al. 1992). It has been suggested that the presence of multiple heterocysts in a maturing cyanobiosis may restrict an efficient transport of photosynthates from the vegetative cells to heterocysts (BERGMAN et al. 1992). In general, however, processes involved in the regulation of the synthesis and activity of nitrogenase are complex and, as yet, poorly understood.

7 The genetics of N_2 fixation in symbiotic cyanobacteria

Studies of the genetics of N_2 fixation in heterocystous cyanobacteria are intimately associated with studies of the process of cellular differentiation in these organisms. However, heterocyst differentiation is not an obligatory prerequisite for aerobic N_2 fixation in all cyanobacteria. Generally in *Anabaena* strains the genetic regulation of N_2 fixation is unusual. For example, as in the free-living *Anabaena* PCC7120, there is no conclusive evidence for the existence of an *ntr*-type regulatory system in this organism. It is possible, after a suitable recovery period, that all cyanobacteria may be capable of synthesising nitrogenase in the presence of ambient O_2.

In some diazotrophic cyanobacteria there is the need to integrate *nif* gene expression with other events (e.g., heterocysts differentiation) and, in these systems, there are almost certainly regulatory signals that are not common to all N_2-fixing organisms (MEEKS et al. 1988, BUIKEMA and HASELKORN 1991, QUISPEL 1992). However, *nif* gene expression in heterocystous cyanobacteria requires the absence of ammonia and oxygen, as in other nitrogen fixing microorganisms.

The difficulty of culturing some cyanobionts (e.g., *Anabaena azollae*) and the lack of a convenient system of genetic analysis have been the major obstacles to the genetic studies of cyanobionts. Nevertheless, in recent years several laboratories have begun to use molecular techniques in order to characterize symbiotic, nitrogen-fixing cyanobacterial strains (FRANCHE and COHEN-BAZIRE 1985, 1987, MEEKS et al. 1988, NIERZWICKI-BAUER and HASELKORN 1986, LINDBLAD et al. 1989, PLAZINSKI et al. 1989, 1990a).

In nitrogen-fixing cyanobacteria, as in all N_2-fixing microorganisms, biological nitrogen fixation is catalyzed by the nitrogenase enzyme complex which is composed of two components; the nitrogenase (called MoFe protein or Protein I) and the nitrogenase reductase (called Fe protein or Protein II), (MORTENSON and THORNELEY 1979). In the most studied free-living cyanobacterium *Anabaena* sp. PCC7120 these components are genetically determined by three genes; *nif*H, *nif*D and *nif*K (RICE et al. 1982). In vegetative cells of *Anabaena* sp. PCC 7120, *nifK* is separated from *nifDH* by an 11-kilobase DNA fragment (RICE et al. 1982), which is excised by site-specific recombination events between directly repeated 11-base pair sequences at each of its ends, and encodes a gene, *xisA*, required for its own excision (LAMMERS et al. 1986). It is also known that a second rearrangement takes place during heterocyst development and leads to the excision of a circular DNA element from vegetative cell DNA near *nif*S (GOLDEN et al. 1985). Recently, a gene controlling heterocyst differentiation, *hetA*, in *Anabaena* sp. PCC7120 has been characterized (BUIKEMA and HASELKORN 1991). In general, a number of the genes that are turned on or off during *Anabaena* heterocyst differentiation have been already cloned and sequenced (reviewed by HASELKORN 1986).

Much less is known about *nif* gene organization in symbiotic cyanobacteria. However, using DNA probes from the free-living strain (*Anabaena* sp. PCC7120) it has been reported that the *nifHDK* genes are contiguous in all cells of *Anabaena azollae* (FRANCHE and COHEN-BAZIRE 1987) and in some fraction of the cells of *Nostoc* sp.7801, a cyanobiont of *Anthoceros punctatus* (MEEKS et al. 1988). These findings imply different organization of the *nif* genes in some symbiotic cyanobacterial strains in comparison to the free-living strains.

FRANCHE and COHEN-BAZIRE (1987) also studied the restriction sites within *nifK, nifD* and *nifH* of four *Anabaena azollae* isolates and have reported that the sites are strongly conserved among the strains with some differences in the *nifH* genes. The same authors also reported a presence of the second copy of *nifH*. In the free-living *Anabaena* sp.PCC7120 the second copy of *nifH* appears to be functional in the sense that it has an open reading frame and is transcribed, and this could also be the case for the symbiotic strains.

In general, genetics of the symbiotic (endophytic) cyanobacteria is still in its infancy and at present, it is difficult to imagine what kind of techniques one should use to better understand the molecular mechanisms of nitrogen fixation processes in cyanobioses.

8 Concluding remarks

In general, processes involved in nitrogen metabolism in symbiotic cyanobacteria are still not well understood. This is mainly because of the difficulty in culturing the cyanobionts and lack of hands-on genetic systems in these organisms. Until these problems are overcome the use of such symbiotic systems in agricultural applications seems unfeasible.

In any plant-microbe interaction, recognition is the primary event that involves signal perception, an activation of regulatory genes followed by the induction of specific biochemical pathways. Since the sites of association (leaf and thallus cavities, coralloid roots, and glands) develop normally in the absence of symbiotic cyanobacteria (PETERS et al. 1986a) it is difficult to assess whether any cyanobionts transmit chemical signals that influence and/or regulate developmental processes in the various host plants. Nevertheless, in lichens, bryophytes and *Azolla* symbioses it has been shown that the hosts synthesize lectins, glycoproteins which specifically recognize sugar residues on the cell surface (reviewed in BERGMAN et al. 1992).

It is now widely accepted that plants synthesise a variety of aromatic compounds (phenolics) in both root and shoot tissues during growth and development (EBEL and HAHLBROCK 1982, HAHLBROCK and GRISEBACH 1979). Recent studies have indicated that plant phenolic compounds induce catabolic genes in many microorganisms as well as genes required to form active associations with plants (PETERS et al. 1986b, STACHEL et al. 1985). It is reasonable to assume that cyanobacteria in association with plants may also respond to specific compounds (inducers) to interact with their respective host plants (SCHMIDT 1991). In fact, a role of trehalose (a common sugar) in cellular communication of various symbiotic systems including endosymbioses has been suggested (MELLOR 1992).

In general, genetic characterization of cyanobioses has hardly begun. Recently, it has been shown that some DNA sequences that are homologous or complementary to *Rhizobium* symbiotic genes are present on indigenous plasmids of *Anabaena azollae* (PLAZINSKI et al. 1991). Therefore, it is very probable that symbiotic cyanobacteria, like other symbiotic nitrogen fixing microorganisms, possess a set of plasmid-borne genes that play an important role in controlling symbiotic processes. As yet such genes have not been identified. To study such aspects of cyanobiosis, the *Gunnera-Nostoc* symbiosis seems to be an especially suitable model system for identifying symbiotic genes, mainly due to its ease in reconstitution.

In addition, there are several other important aspects of the cyanobacterial-plant symbiosis which should be given priority in future experimental work. For instance, processes of regulation and development of the increased number of heterocysts, and translocation of photosynthates to cyanobionts are still unknown. Moreover, the cyanobiont's inability to assimilate all the ammonia produced during nitrogen fixation process is worth investigating.

Studying the mechanisms of regulatory control in cyanobioses will certainly improve our understanding of these unusual associations, and this seems to be more feasible now with rapid progress being made in molecular biology. The knowledge gained from such studies will certainly be extrapolated into other, more agriculturally important plant-microbe associations.

Acknowledgments

I thank Professor R.W. RIDGE for his critical review of this manuscript.

References

Ahmadjian V (1982) Algal/fungal symbioses In: Round FE, Chapman V (eds) Elsevier, Amsterdam, pp 179-233

Basilier K, Granhall U (1978) Nitrogen fixation in wet minerotrophic moss communities of the subarctic mire. Oikos 31:236-246

Becking JH (1976) Nitrogen fixation in some natural ecosystems in Indonesia. In: Nutman PS (ed) Symbiotic nitrogen fixation in plants. Cambridge University Press, Cambridge, pp 539-550

Bergman B, Rai AN (1989) The *Nostoc-Nephroma* symbiosis: localization, distribution pattern and levels of key proteins involved in nitrogen and carbon metabolism of the cyanobiont. Physiol Plant 77:216-224

Bergman B, Rai AN, Johansson C, Soderback E (1992) Cyanobacterial-plant symbioses. Symbiosis 14:61-81

Bonnett HT, Silvester WB (1981) Specificity in the *Gunnera-Nostoc* endosymbiosis. New Phytol 89:121-128

Bothe H, Nelles H, Hager K-P, Papen H, Neuer G (1984) Physiology and biochemistry of N_2 fixation by cyanobacteria. In: Veeger C and Newton WE (eds) Advances in nitrogen fixation research. The Hague and Wageningen: Nijhoff, Junk, Pudoc pp 199-210

Braun-Howland EB, Nierzwicki-Bauer SA (1990) *Azolla-Anabaena* symbiosis: biochemistry, physiology, ultrastructure and molecular biology. In: Rai AN (ed) Handbook of symbiotic cyanobacteria. CRC Press, Boca Raton, Florida, pp 65-117

Bubrick P, Galun M, Ben-Yaacou M, Frensdorff A (1982) Antigenic similarities and differences between symbiotic and cultured phycobionts from the lichen *Xanthoria parietina*. FEMS Microbiol Lett 13:435-438

Buikema WJ, Haselkorn R (1991) Characterization of a gene controlling heterocyst differentiation in the cyanobacterium *Anabaena* 7120. Genes & Development 5:321-330

Burris RH (1979) Inhibition. In: Hardy RWF and Burns RC (eds) A treatise on dinitrogen fixation, section 2, Wiley, New York, pp 569-604

Caiola MG, Forni C, Castagnola M (1988) Bacteria in the *Azolla-Anabaena* association. Symbiosis 5:185-198

Calvert HE, Peters GA (1981) The *Azolla-Anabaena* relationship. IX. Morphological analysis of leaf cavity hair populations. New Phytol 89:327-335

Carrapico F (1991) Are bacteria the third partner of the *Azolla-Anabaena* symbiosis? Plant Soil 137:157-160

Dixon ROD, Wheeler CT (1986) Nitrogen fixation in plants. Chapman and Hall, Blackie, New York, pp 101-103

Duckett JG, Toth R, Soni SL (1975) An ultrastructural study of the *Azolla, Anabaena azollae* relationship. New Phytol 75:111-118

Ebel J, Hahlbrock K (1982) Biosynthesis. In: Harborne JB, Mabry TJ (eds) The flavonoids, advances in research. Chapman and Hall, New York, pp 641-679

Fogg GE, Stewart WDP, Fay P, Walsby AE (ed) (1973) The blue-green algae. Academic Press, London, pp 221-245

Franche C, Cohen-Bazire G (1985) The structural *nif* genes of four symbiotic *Anabaena azollae* show a highly conserved physical arrangement. Plant Sci 39:125-131

Franche C, Cohen-Bazire G (1987) Evolutionary divergence in the *nifH D K* gene region among nine symbiotic *Anabaena azollae* and between *Anabaena azollae* and some free-living heterocystous cyanobacteria. Symbiosis 3:159-178

Gallon JR, Chaplin AE (ed) (1987) An introduction to nitrogen fixation. Cassell, London

Galun M, Bubrick P (1984) Physiological interactions between the partners of the lichen symbiosis. In: Linskens HF, Heslop-Harrison J (eds) Cellular interactions, Encyclopedia of plant physiology, vol 17. Springer Verlag, Berlin, pp 362-401

Gates JE, Fisher RW, Goggin TW, Azrolan NI (1980) Antigenic differences between *Anabaena azollae* fresh from the *Azolla* fern leaf cavity and free-living cyanobacteria. Arch Microbiol 128:126-129

Gebhardt JS, Nierzwicki-Bauer SA (1991) Identification of a common cyanobacterial symbiont associated with *Azolla* spp. through molecular and morphological characterization of free-living and symbiotic cyanobacteria. Appl Environ Microbiol 57:2141-2146

Golden JW, Robinson SJ, Haselkorn R (1985) Rearrangement of nitrogen fixation genes during heterocyst differentiation in the cyanobacterium *Anabaena*. Nature 314:419-423

Granhall U, Hofsten AV (1976) Nitrogenase activity in relation to intracellular organisms in *Sphagnum* mosses. Physiol Plant 36:88-94

Grobbelaar N, Scott WE, Hattingh W, Marshall J (1987) The identification of the coralloid root endophytes of the southern African cycads and the ability of the isolates to fix nitrogen. S Afr J Bot 53:111-118

Gunning BES, Pate JS (1969) "Transfer cells": plant cells with wall ingrowths specialized in relation to short distance transport of solutes-their occurrence, structure, and development. Protoplasma 68:107-111

Hahlbrock K, Grisebach H (1979) Enzymatic control in the biosynthesis of lignin and flavonoids. Annu. Rev. Plant Physiol. 30:105-130

Halliday J, Pate JS (1976) Symbiotic nitrogen fixation by coralloid roots of the cycad *Macrozamia riedlii*: physiological characteristics and ecological significance. Aust J Plant Physiol 3:349-358

Haselkorn R (1978) Heterocysts. Annu Rev Plant Physiol 29:319-344

Haselkorn R (1986) Organization of the genes for nitrogen fixation in photosynthetic bacteria and cyanobacteria. Annu Rev Microbiol 40:525-547

Hitch CJB, Millbank JW (1975) Nitrogen metabolism in lichens VII; nitrogenase activity and heterocyst frequency in lichens with blue-green phycobionts. New Phytol 75:239-244

Hom SSM, Hennecke H, Shanmugam KT (1980) Regulation of nitrogenase biosynthesis in *Klebsiella pneumoniae*: effect of nitrate. J Gen Microbiol 117:169-179

Isaac S (1992) (ed) Fungal-plant interactions. Chapman and Hall, London, New York, Tokyo, Melbourne, Madras, pp 298-326

Johansson C, Bergman B (1992) Early events during the establishment of the *Gunnera/Nostoc* symbiosis. Planta 188:403-413

Joseph CM, Meeks JC (1987) Regulation of expression of glutamine synthetase in a symbiotic *Nostoc* strain associated with *Anthoceros punctatus*. J Bacteriol 169:2471-2475

Kikuchi M, Watanabe I, Haws LD (1984) Economic evaluation of *Azolla* use in rice production. In: The International Rice Research Institute (ed) Organic matter and rice. P.O. Box 933, Manila, Philippines, pp 569-592

Lammers PJ, Golden JW, Haselkorn R (1986) Identification and sequence of a gene required for a developmentally regulated DNA excision in *Anabaena*. Cell 44:905-911

Lin C, Watanabe I (1988) A new method for obtaining *Anabaena-free Azolla*. New Phytol 108:341-344

Lin C, Watanabe I, Liu C-C, Zheng D-Y, Tang LF (1988) Re-establishment of symbiosis to *Anabaena-free Azolla*. In: Bothe H, de Bruijn FJ, Newton WE (eds), Nitrogen fixation: hundred years after. Gustav Fisher, Stuttgart, pp 223-227

Lindblad P, Bergman B (1986) Glutamine synthetase: activity and localization in cyanobacteria of the cycads *Cycas revoluta* and *Zamia skinneri*. Planta 169:1-7

Lindblad P, and Bergman B (1990) The cycad-cyanobacterial symbiosis. In: Rai AN (ed) Handbook of symbiotic cyanobacteria. CRC Press, Boca Raton, Florida, pp 43-63

Lindblad P, Haselkorn R, Bergman B, Nierzwicki-Bauer SA (1989) Comparison of DNA restriction fragment length polymorphisms of *Nostoc* strains in and from cycads. Arc Microbiol 152:20-24

Lindblad P, Bergman B, Nierzwicki-Bauer SA (1991) Immunocytochemical localization of nitrogenase in bacteria symbiotically associated with *Azolla* spp. Appl Environ Microbiol 57:3637-3640

Liu C-C, Chen Y, Tang L, Zheng Q, Song T, Chen M, Li Y, Lin T (1989) Study on preparation and application of monoclonal antibodies to *Anabaena azollae*. Science in China (Ser B) 32:562-569

Lumpkin TA, Plucknett C (1982) *Azolla* as a green manure: use and management in crop production. West View Press, Boulder, pp 112-116

Meeks JC, Enderlin CS, Joseph CM, Chapman JS, Lollar MWL (1985a) Fixation of $[^{13}N]N_2$ and transfer of fixed nitrogen in the *Anthoceros-Nostoc* symbiotic association. Planta 164:406-414

Meeks JC, Steinberg N, Joseph CM, Enderlin CS, Jorgensen PA, Peters GA (1985b) Assimilation of exogenous and dinitrogen-derived $^{13}NH_4^+$ by *Anabaena azollae* separated from *Azolla caroliniana* Willd. Arch Microbiol 142:229-233

Meeks JC, Joseph CM, Haselkorn R (1988) Organization of the *nif* genes in cyanobacteria in symbiotic association with *Azolla* and *Anthoceros*. Arch Microbiol 150:61-71

Meeks JC (1990) Cyanobacterial bryophyte associations. In: Rai AN (ed) Handbook of symbiotic cyanobacteria. CRC Press, Boca Raton, Florida, pp 43-63

Mellor R (1992) Is trehalose a symbiotic determinant in symbioses between higher plants and microorganisms? Symbiosis 14:113-129

Milicia F, Favilli F (1992) *Azolla* symbiotic system's application as biofertilizer for green garden crops. Symbiosis 14:495-500

Millbank JW (1974) Associations with blue-green algae. In: Quispel A (ed) The biology of nitrogen fixation. American Elsevier, New York, pp 238-264

Millbank JW (1984) Nitrogen fixation by lichens. In: Subba Rao NS (ed) Current developments in biological nitrogen fixation. Edward Arnold, London, pp 197-218

Moore AW (1969) *Azolla*: biology and agronomic significance. Bot Rev 35:17-34

Mortenson LE, Thorneley RNF (1979) Structure and function of nitrogenase. Annu Rev Biochem 48:387-411

Neumann D, von Ackermann M, Jacob F (1970) Zur Feinstruktur der endophytischen Cyanophyceen von *Gunnera chilensis* Lam. Biochem Physiol Pflanz 161:483-498

Newton JW, Herman AI (1979) Isolation of cyanobacteria from the aquatic fern, *Azolla*. Arch Microbiol 120:161-165

Nierzwicki-Bauer SA, Aulfinger H (1990) Ultrastructural characterization of eubacteria residing within leaf cavities of symbiotic and cyanobiont-free *Azolla mexicana*. Curr Microbiol 21:1-6

Nierzwicki-Bauer SA, Haselkorn R (1986) Differences in mRNA levels in *Anabaena* living freely or in symbiotic association with *Azolla*. EMBO J 5:29-35

Orr J, Haselkorn R (1982) Regulation of glutamine synthetase activity and synthesis in free-living and symbiotic *Anabaena* spp. J Bacteriol 152:626-635

Pankow H, Martens B (1964) Uber *Nostoc sphaericum* Vauch. Arch Mikrobiol 48:203-212

Peters GA (1977) The *Azolla-Anabaena* symbiosis. In: Hollaender A (ed) Genetic engineering and nitrogen fixation. Plenum, New York, pp 231-258

Peters GA, Ray TB, Mayne BC, Toia RE, Jr (1980) *Azolla-Anabaena* association: morphological and physiological studies. In: Newton WE, Orme-Johnson WH (eds) Nitrogen fixation. University Park Press, Baltimore, pp 293-309

Peters GA, Kaplan D, Meeks JC, Buzby KM, Marsh BH, Corbin JL (1985) Aspects of carbon and nitrogen exchange in the *Azolla-Anabaena* symbiosis. In: Ludden PW, Burris JE (eds) Nitrogen fixation and CO_2 metabolism. Elsevier Science, New York, pp 213- 222

Peters GA, Toia RE Jr, Calvert HE, Marsh BH (1986a) Lichens to *Gunnera* - with emphasis on *Azolla*. Plant and Soil 90:17-34

Peters GA, Meeks JC (1989) The *Azolla-Anabaena* symbiosis: basic biology. Annu Rev Plant Physiol Plant Mol Biol 40:193-210

Peters NK, Frost JW, Long SR (1986b) A plant flavone luteolin, induces expression of *Rhizobium meliloti* nodulation genes. Science 233:977-980

Petro MJ, Gates JE (1987) Distribution of *Arthrobacter* sp. in the leaf cavities of four species of the N_2-fixing *Azolla* fern. Symbiosis 3:41-48

Plazinski J, Zheng Q, Taylor R, Rolfe BG, Gunning BES (1989) Use of DNA/DNA hybridization techniques to authenticate the production of new *Azolla-Anabaena* symbiotic association. FEMS Microbiol Lett 65:199-204

Plazinski J (1990) The *Azolla-Anabaena* symbiosis. In: Gresshoff PM (ed) Molecular biology of symbiotic nitrogen fixation. CRC Press, Boca Raton, Florida, pp 51-74

Plazinski J, Zheng Q, Taylor R, Croft L, Rolfe BG, Gunning BES (1990a) DNA probes show genetic variation in cyanobacterial symbionts of the *Azolla* fern and a closer relationship to free-living *Nostoc* strains than to free-living *Anabaena* strains. Appl Environ Microbiol 56:1263-1270

Plazinski J, Taylor R, Shaw W, Croft L, Rolfe BG, Gunning BES (1990b) Isolation of *Agrobacterium* sp., strain from the *Azolla* leaf cavity. FEMS Microbiol Lett 70:55-60

Plazinski J, Croft L, Taylor R, Zheng Q, Rolfe BG, Gunning BES (1991) Indigenous plasmids in *Anabaena azollae;* their taxonomic distribution and existence of regions of homology with symbiotic genes of *Rhizobium.* Can J Microbiol 37:171-181

Quispel A (1992) A search for signals in endophytic microorganisms. In: Verma DPS (ed) Molecular signals in plant-microbe communications. CRC Press, Boca Raton, Ann Arbor, London, pp 471-491

Rai AN (1990) Cyanobacterial-fungal symbioses: the cyanolichens. In: Rai AN (ed) Handbook of Symbiotic cyanobacteria. CRC Press, Boca Raton, Florida, pp 9-41

Ray TB, Peters GA, Toia RE Jr, Mayne BC (1978) *Azolla-Anabaena* relationship VII. Distribution of ammonia-assimilating enzymes, protein, and chlorophyll between host and symbiont. Plant Physiol 62:463-467

Ray TB, Mayne BC, Peters GA, Toia RE, Jr (1979) *Azolla-Anabaena* relationship. VIII. Photosynthetic characterization of the association and individual partners. Plant Physiol 64:791-795

Rice D, Mazur BJ, Haselkorn R (1982) Isolation and physical mapping of nitrogen fixation genes from the cyanobacterium *Anabaena* PCC7120. J Biol Chem 257:13157-13163

Rippka R, Deruelles J, Waterbury JB, Herdman M, Stanier RY (1979) Generic assignments, strain histories and properties of pure cultures of cyanobacteria. J Gen Microbiol 111:1-61

Rodgers GA, Stewart WDP (1977) The cyanophyte-hepatic symbiosis. I. Morphology and physiology, New Phytol 78:441-458

Schmidt H (1991) Licht-und elektronenmikroskopische Untersuchungen zum Infektionsablauf der *Gunnera-Nostoc*-Symbiose. Dissertationes Botanica, Band 177. Cramer J, Gebruder Borntraeger, Berlin Stuttgart

Silvester WB (1976) Endophytic adaptation in *Gunnera-Nostoc* symbiosis. In: Nutman PS (ed) Symbiotic nitrogen fixation in plants. Cambridge University Press, Cambridge, pp 521-538

Silvester WB, McNamara PJ (1976) The infection process and ultrastructure of the *Gunnera-Nostoc* symbiosis. New Phytol 77:135-141

Simon RD (1971) Cyanophycin granules from the blue-green alga *Anabaena cylindrica*: a reserve material consisting of copolymers of aspartic acid and arginine. Proc Natl Acad Sci USA 68:265-269

Smith SE, Smith FA (1990) Structure and function of the interface in biotrophic symbiosis and their relation to nutrient transport. New Phytol 114:1-6

Soderback E, Lindblad P, Bergman B (1990) Developmental patterns related to nitrogen fixation in the *Nostoc-Gunnera magellanica* Lam. symbiosis. Planta 182:355-362

Stachel SE, Messeus E, von Montagu M, Zambryski P (1985) Identification of the signal molecules produced by wounded plant cells that activate T-DNA transfer in *Agrobacterium tumefaciens*. Nature 318:624-629

Stanier RY, Cohen-Bazire G (1977) Phototrophic prokaryotes:the cyanobacteria. Annu Rev Microbiol 31:225-274

Stewart WDP, Rodgers GA (1977) The cyanophyte-hepatic symbiosis. II. Nitrogen fixation and interchange of nitrogen and carbon. New Phytol 78:459-471

Stewart WDP (1980) Some aspects of structure and function in N_2-fixing cyanobacteria. Annu Rev Microbiol 34:497-536

Stewart WDP, Rowell P (1975) Effects of L-methionine-DL-sulphoximine on the assimilation of newly fixed NH_3, acetylene reduction and heterocyst production in *Anabaena cylindrica*. Biochem Biophys Res Commun 65:846-851

Stewart WDP, Rowell P, Rai AN (1980) Symbiotic nitrogen-fixing cyanobacteria. In: Stewart WDP, Gallon JR (eds) Nitrogen fixation. Academic Press, New York, pp 239-277

Stewart WDP, Rowell P, Rai AN (1983) Cyanobacteria- eukaryotic plant symbioses. Ann Microbiol (Inst Pasteur), 134B:205-228

Stewart WDP, Rowell P, Cossar JD, Kerby NW (1985) Physiological studies on N_2-fixing cyanobacteria. In: Ludden PW and Burris JE (eds). Nitrogen fixation and CO_2 metabolism. Elsevier, New York, pp 269-279

Strasburger E (1873) Uber *Azolla*. Jena: Hermann Dabis Verlag, p 86

Thomas JH, Mullineaux PM, Cronshaw AD, Chaplin AE, Gallon JR (1982) The effects of structural analogues of amino acids on ammonium assimilation and acetylene reduction (nitrogen fixation) in *Gloeocapsa* (*Gloeothece*) sp. CCAP 1430/3. J Gen Microbiol 128:885-893

Tredici MR, Margheri MC, Giovanetti L, de Philippis R, Vincenzini M (1988) Heterotrophic and diazotrophic growth of *Nostoc* sp. from *Cycas circinalis*. Plant Soil 110:199-203

Tyagi VVS, Ray TB, Mayne BC, Peters GA (1981) The *Azolla-Anabaena azollae* relationship XI. Phycobiliproteins in the action spectrum for nitrogenase-catalyzed acetylene reduction. Plant Physiol 68:1479-1484

Wolk CP, Thomas J, Shaffer PW, Austin SM, Galonsky A (1976) Pathway of nitrogen metabolism after fixation of ^{13}N-labeled nitrogen gas by the cyanobacterium, *Anabaena cylindrica*. J Biol Chem 251 5027-5033

Wallace WH, Gates JE (1986) Identification of Eubacteria isolated from leaf cavities of four species of the N_2-fixing *Azolla* fern as *Arthrobacter* Conn and Dimmick. Appl Environ Microbiol 52:425-429

Walsby AE, Nichols BW (1969) Lipid composition of heterocysts. Nature 221:673

Watanabe I (1977) *Azolla* utilization in rice culture. Int Rice Res Newslett 2:3-8

Watanabe I, Roger PA (1984) Nitrogen fixation in wetland rice fields. In: Rao NSS (ed), Current developments in biological nitrogen fixation. Edward Arnold, London, pp 237-276

Wilcox M (1970) One-dimensional pattern found in blue-green algae. Nature 228-686

Youngman RJ (1984) Oxygen activation: is the hydroxyl radical always biologically relevant? TIBS 9:280-282

Zimmerman WJ, Rosen BH, Lumpkin TA (1989) Enzymatic, lectin, and morphological characterization and classification of presumptive cyanobionts from *Azolla* Lam. New Phytol 113:497-503

Interactions between Carbon and Nitrogen Metabolism

CATALINA LARA and MIGUEL G. GUERRERO

Instituto de Bioquímica Vegetal y Fotosíntesis, Universidad de Sevilla-CSIC, Apdo, 1113, 41080-Sevilla, Spain

1 Introduction

As in other photosynthetic organisms, carbon and nitrogen are two of the most abundant elements in the cyanobacteria, accounting for about 50% and 10%, respectively, of their dry weight. The major sources of carbon and nitrogen available in most environments to the generality of cyanobacteria are carbon dioxide and nitrate, which are assimilated photosynthetically. CO_2 fixation and nitrate assimilation actually represent the two major processes in the photosynthetic metabolism of cyanobacteria. In addition to nitrate, cyanobacteria can use ammonium as primary nitrogen source and the diazotrophic strains can utilize atmospheric nitrogen. The metabolic and regulatory integration of photosynthetic carbon and nitrogen assimilation is, therefore, a main issue in cyanobacterial physiology.

In this chapter, we attempt to update and summarize the present status of knowledge of the various and complex interactions between nitrogen and carbon metabolism in cyanobacteria, especially with respect to two different aspects: the incidence of inorganic nitrogen assimilation on carbon metabolism, and the metabolic and regulatory role of CO_2 fixation on nitrogen assimilation. This review namely deals with studies on the subject carried out in our laboratory, although relevant contributions by other authors are considered.

2 Assimilation of inorganic nitrogen and carbon

For the present purpose, inorganic nitrogen assimilation will be considered as the conversion of an external inorganic nitrogen source into the α-amino group of an amino acid. Usually, and provided that light and CO_2 are not limiting, pH is kept constant and the nitrogen source is present in adequate concentrations, the rate of inorganic nitrogen assimilation, as well as the growth rate, for a given cyanobacterium is much the same regardless the nature of the nitrogen source (OHMORI and HATTORI 1972, FOGG et al. 1973, BOTTOMLEY et al. 1979). As only ammonium-nitrogen is directly incorporated into amino acids, nitrate or dinitrogen have to be reduced to ammonium, prior to its incorporation into amino acids, by the assimilatory pathways depicted in Fig. 1 and described in detail in other chapters of this volume.

Although simultaneous assimilation of two nitrogen sources can take place, at least transitorily, the general rule is that ammonium, when available, is assimilated in preference to nitrate, the latter being, in turn, preferred to N_2. This behaviour, which is the manifestation of regulatory interactions between the different assimilatory pathways involved (nitrogen control; GUERRERO and LARA 1987), is in agreement with the logics of cell economy, since the energy requirements for the assimilation of one nitrogen equivalent from the different nitrogen forms to the level of the amino-nitrogen of glutamate increase in the order: ammonium (2

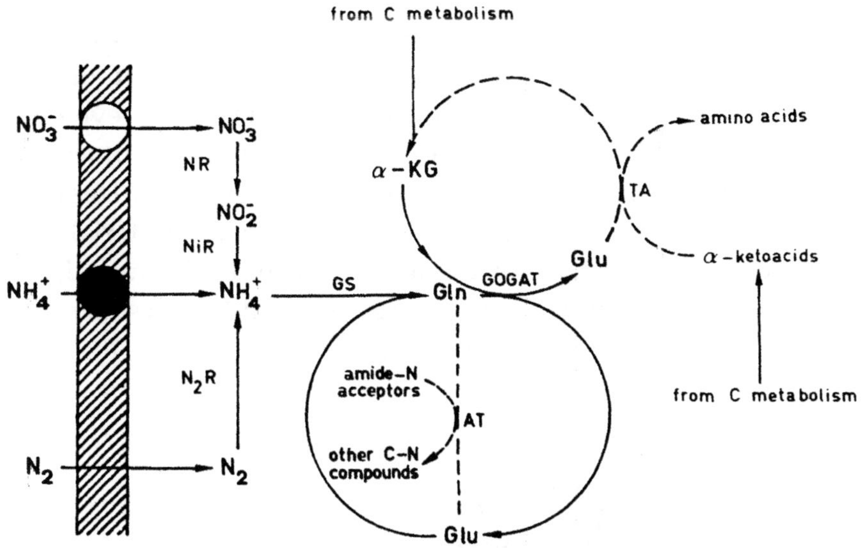

Fig. 1. Pathways of nitrate, ammonium and dinitrogen assimilation in cyanobacteria. The hatched area on the left represents the plasma membrane. Dotted lines indicate transamination and transamidation reactions. AT, glutamine (Gln)-amide transferase activities; GOGAT, glutamate (Glu) synthase; GS, glutamine synthetase; NR, ferredoxin-nitrate reductase; NiR, ferredoxin-nitrite reductase; N_2R, dinitrogen reductase; TA, Glu-amino transferase activities. Redrawn from GUERRERO and LARA (1987).

reducing equivalents and 1 ATP, plus the energy used for transport), nitrate (10 reducing equivalents and 1 ATP, plus the energy used for transport), and dinitrogen (6 reducing equivalents and a minimum of 7 ATP).

As for inorganic carbon, cyanobacteria, which can transport either CO_2 or bicarbonate from the medium (MILLER et al. 1990), assimilate CO_2 by the reductive pentose phosphate cycle. Hexosephosphate are used as substrates for the biosynthesis of glycogen, the major reserve carbohydrate, via ADP-glucose pyrophosphorylase. Triose-phosphate, in turn, may flow through the glycolytic and the tricarboxylic acid pathways to provide 2-oxoacids which will eventually be used as amino acceptors for amino acid biosynthesis. Cyanobacteria have an incomplete tricarboxylic acid pathway lacking both 2-oxoglutarate dehydrogenase and succinyl-CoA synthetase activities (STANIER and COHEN-BAZIRE 1977). The operativity of the pathway is mainly biosynthetic in nature, and yields 2-oxoglutarate on one end, and succinate, on the other end (Fig. 2). It requires the anaplerotic function of phosphoenolpyruvate carboxylase, to provide not only the oxaloacetate needed for the synthesis of citrate and, eventually, 2-oxoglutarate, but also for that used in the synthesis of aspartate and succinate, a precursor for chlorophyll biosynthesis. This crucial role of phosphoenol-pyruvate carboxylase in cyanobacterial carbon metabolism is the basis for the high capacity of the so-called "C_4 carboxylation" observed in these organisms, although *sensu strictus* they have a C_3 -type of CO_2 fixation.

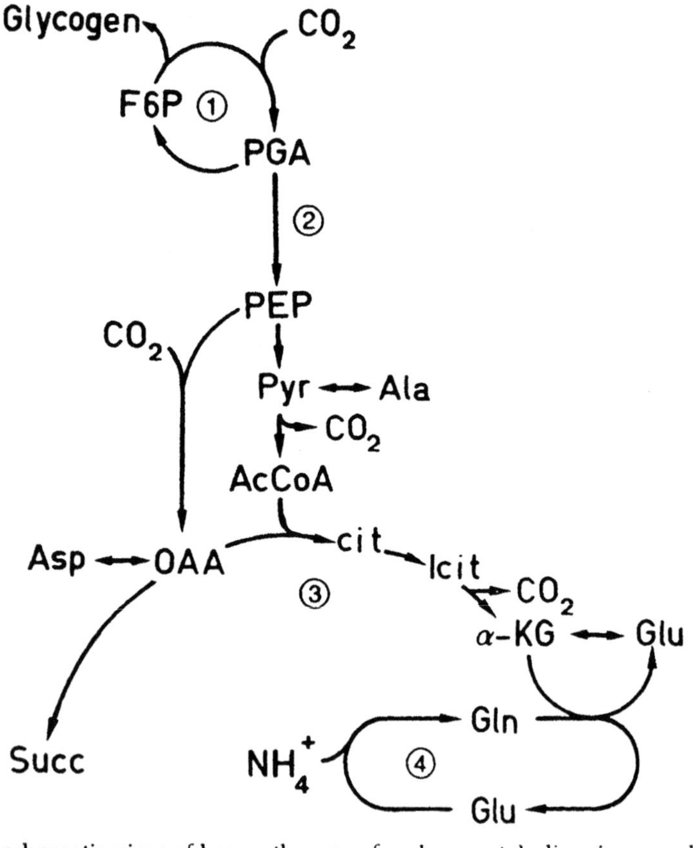

Fig. 2. A schematic view of key pathways of carbon metabolism in cyanobacteria. Encircled numbers represent: 1, reductive pentose-phosphate cycle; 2, glycolysis; 3, incomplete tricarboxylic acid pathway; and 4, GS-GOGAT pathway. F6P, fructose 6-phosphate; PGA, phosphoglycerate; PEP, phosphoenolpyruvate; Pyr, pyruvate; Ala, alanine; AcCoA, acetyl coenzyme A; OAA, oxaloacetate; Asp, aspartate; Succ, succinate; cit, citrate; Icit, isocitrate; α-KG, α-ketoglutatate. Redrawn from CORONIL et al. (1993).

It is evident that in the assimilation of any form of inorganic nitrogen, an obligate requirement for carbon skeletons, mostly 2-oxoacids, arises at the level of ammonium assimilation by the glutamine synthetase-glutamate synthase pathway and related transamination and transamidation reactions (Fig. 1). This requirement will always exist, regardless the nature of the nitrogen source assimilated, and will lead in all cases to a diversion of carbon flow from carbohydrate biosynthesis to amino acid biosynthesis (discussed below in section 3). Its extent can be evaluated by measuring the dependence of ammonium assimilation upon CO_2 fixation. Thus, sustained ammonium assimilation in *Anacystis* cells is dependent upon the availability of CO_2 (BOUSSIBA et al. 1984), and cannot proceed when CO_2 fixation has been specifically inhibited by treatment with D,L-glyceraldehyde (LARA and ROMERO 1986). In addition to this, when nitrate is the nitrogen source, assimilatory power is needed for nitrate reduction to ammonium. The reducing power demand, which, as discussed in section 4, is directly fulfilled by the photosynthetic apparatus, may determine competition with CO_2 fixation for reducing equivalents. Also, when N_2 is the nitrogen source, carbohydrate oxidation can be the immediate source of the assimilatory power required for N_2 fixation (discussed in section 5), thus increasing the carbon requirement of this assimilatory process. Finally, regulatory interactions between carbon metabolism and nitrate and dinitrogen assimilation exist, determining the rate of nitrogen assimilation according to the carbon/nitrogen status of the cell.

3 Incidence of ammonium assimilation on CO_2 fixation

The incorporation of ammonium to carbon skeletons may affect to both the rate of CO_2 fixation and the distribution of newly fixed carbon between metabolite fractions. These ammonium effects can be analyzed and evaluated by studying the situation in cells actively fixing CO_2 in the absence and presence of ammonium. When other inorganic nitrogen form is assimilated, the ammonium effects will also exists, superimposed to other metabolic and regulatory interactions (discussed in the next sections), and should also be taken into account.

Studies carried out with the unicellular cyanobacterium *Anacystis nidulans* show that in nitrate-grown cells exposed to ammonium, the rate of CO_2 fixation is higher than in those cells assayed in the absence of inorganic nitrogen, this effect being particularly manifest at high light intensities (ROMERO and LARA 1987, GARCÍA-GONZÁLEZ et al. 1992, CORONIL et al. 1993). This ammonium-induced stimulation of CO_2 fixation appears to be related to the fact that, as in other photosynthetic organisms (KANAZAWA et al. 1972, PAUL et al. 1978, WOO and CANVIN 1980a,b), ammonium incorporation to carbon skeletons diverts the carbon flow to organic and amino acid biosynthesis at the expense of reserve carbohydrate biosynthesis (GARCÍA-GONZÁLEZ et al. 1992, CORONIL et al. 1993).

Thus, in *A. nidulans* cells supplied with saturating $^{14}CO_2$, addition of ammonium increases the percentage of recently fixed carbon incorporated into acid-soluble metabolites by about 3-fold. Within this fraction, the labeling of sugar-phosphates is practically unchanged, that of organic acids increases, and the most dramatic changes are observed in the percentage of carbon incorporated into amino acids, which increases about 11-fold under light-intensities saturating for both ammonium assimilation and CO_2 fixation, and about 4-fold under light-intensities saturating for ammonium assimilation but only half-saturating for CO_2 fixation. In contrast, the pool size of total amino acids increases less than 2-fold (CORONIL et al. 1993). As carbon incorporation into amino acids reflects *de novo* amino acid biosynthesis, whereas their pool size reflects the balance between generation and utilization, the turnover of each amino acid is, therefore, reflected in its specific-radioactivity value (i.e., the ratio between the g atom of ^{14}C incorporated per mol of amino acid). At the two light-intensities selected, ammonium assimilation dramatically accelerated the turnover of most amino acids. This is particularly evident for the case of glutamate, the main α-amino donor in transamination reactions, whose specific radioactivity increased 17-fold and 6-fold at high and low light-intensity, respectively. As this carbon diversion induced by ammonium assimilation is more marked under fully saturating light-intensity, it appears that under conditions allowing maximal operativity of the glutamine synthetase-glutamate synthase pathway, the synthesis of glutamate and the subsequently increased capacity for transamination with different 2-oxoacids represent an effective sink for newly fixed carbon (CORONIL et al. 1993).

On the basis that the transamination reactions involving the 2-oxoacids pyruvate (Pyr), oxaloacetate (OAA) and 2-oxoglutarate (2OG) and the corresponding amino acids Ala, Asp and Glu are close to equilibrium, it can safely be assumed that specific-radioactivity ratios for Asp/Ala, Glu/Ala and Glu/Asp are equivalent to those for OAA/Pyr, 2OG/Pyr and 2OG/OAA, respectively. The specific radioactivity ratio for Asp/Ala can be taken, therefore, as an index of the flow of newly fixed carbon to OAA, via phosphoenolpyruvate carboxylase, relative to the carbon flow to Pyr, via glycolysis. Similarly, the respective Glu/Ala ratio reflects the flow of newly fixed carbon to 2-oxoglutarate relative to that through Pyr, and the Glu/Asp ratio, the carbon flow to 2OG relative to that to OAA. As illustrated in Table 1, ammonium assimilation induces a dramatic increase in the Glu/Ala and the Glu/Asp ratios, especially at saturating light, consistent with increased carbon flow through Pyr to 2OG. That would require not only stimulation of Pyr kinase, as has also been proposed for N-limited green algae (BASSHAM et al. 1981, ELRIFI and TURPIN 1987, SMITH et al. 1989), but also increased activity of Pyr dehydrogenase and/or citrate synthase (CORONIL et al. 1993).

Table 1. Ratios of specific radioactivity for amino acids in *A. nidulans* cells after 15 min of $^{14}CO_2$ fixation under light-saturating conditions. From CORONIL et al. (1993)

	Minus N	Plus NO_3^-	Plus NH_4^+
Asp/Ala	1.9	3.5	1.8
Glu/Ala	0.3	3.0	3.6
Glu/Asp	0.2	0.9	2.1

The diversion of carbon flow in N-sufficient *A. nidulans* cells has no negative effects on the operativity of the Calvin cycle and, in fact, it is produced at the expense of carbon flow towards glycogen biosynthesis. Under conditions of ammonium assimilation, the percentage of newly fixed carbon incorporated into glycogen decreases by about 50%, under low light intensity, and by about 30% at saturating light intensity. Ammonium assimilation also promotes a significant decrease in net glycogen accumulation, which amounts to 80% and 20% at low and high light intensities, respectively. Interestingly, at low light intensity ammonium induces a 2-fold increase in the specific radioactivity of the glycogen pool, indicating acceleration of glycogen turnover (Table 2). As net glycogen accumulation is severely restricted under these conditions, an ammonium-induced stimulation of glycogen degradation is inferred. At high light-intensity, in contrast, the specific radioactivity of glycogen is slightly lower under ammonium assimilation conditions, indicating that the major ammonium effect is to decrease carbon flow towards glycogen biosynthesis (GARCÍA-GONZÁLEZ et al. 1992).

Table 2. Specific ^{14}C radioactivity of the glycogen accumulated in *A. nidulans* cells after 20 min of $^{14}CO_2$ fixation. From GARCÍA-GONZÁLEZ et al. (1992).

Nitrogen source	Specific radioactivity (g atom ^{14}C mol^{-1} glucose)	
	Half saturating light	Saturating light
None	1.7	2.4
Ammonium	3.4	1.9

The mechanism whereby the activity of key enzymes of glycogen synthesis and degradation responds to ammonium assimilation is still an open question. ADP-glucose pyrophosphorylase, a regulatory enzyme of glycogen biosynthesis, has been shown in *A. nidulans* to be inhibited by inorganic phosphate, with an $I_{0.5}$ value of 72 μM (LEVY and PREISS 1976). Intracellular levels of inorganic phosphate could probably increase under conditions of ammonium assimilation, since acceleration of carbon flow to amino acid biosynthesis would contribute to inorganic phosphate release from sugar-phosphate. Under these conditions, inhibition of ADP-glucose pyrophosphorylase activity by P_i would decrease carbon flow to glycogen favouring the diversion of carbon flow to amino acid biosynthesis, while faster inorganic phosphate recycling, which is an intrinsic factor limiting CO_2 fixation (WALKER and SIVAK 1986), would allow higher CO_2 fixation rates, as observed under conditions of ammonium assimilation (ROMERO and LARA 1987, GARCÍA-GONZÁLEZ et al. 1992, CORONIL et al. 1993).

The ammonium effect on newly fixed carbon distribution can also be observed when other nitrogen source, such as nitrate or dinitrogen, is assimilated, although the extent of the carbon shift is much more moderate. Thus, under conditions of nitrate assimilation, diversion of carbon flow from glycogen to amino acid biosynthesis in *A. nidulans* cells occurs, superimposed to other effects. Interestingly, no stimulation of glycogen degradation has been observed under conditions of nitrate assimilation, and the acceleration of amino acid turnover is less dramatic than that elicited by ammonium assimilation, indicating that the carbon skeletons demand for amino acid biosynthesis during nitrate assimilation is not as strong as that observed during ammonium assimilation (GARCIA-GONZÁLEZ et al. 1992, CORONIL et al. 1993). Also, in N-deficient *Anabaena cylindrica* cells, exposure to N_2 promotes a dramatic increase in carbon flow to amino acids as a consequence of nitrogen fixation and incorporation of ammonium to carbon skeletons (LAWRIE et al. 1976). In this latter case, the N-deficiency of the cells, and the subsequently high C/N ratio, determines a transient increase in the N_2 fixation capacity that contributes to the magnification of the carbon shift.

4 Interactions between nitrate assimilation and CO_2 fixation

In addition to the ammonium effects described above, under conditions of nitrate assimilation two different levels of interaction with CO_2 fixation appear, namely, a competition for assimilatory power between both processes under certain conditions and a carbon-control of nitrate assimilation exerted on the transport of nitrate into the cell.

4.1 Competition for assimilatory power

In cyanobacteria, as in other photosynthetic cells and tissues, the assimilatory power required for nitrate assimilation is directly provided by the photosynthetic apparatus (see LARA 1992, for a recent review).

Quantitative photochemical conversion of nitrate to ammonium has been demonstrated by measuring nitrate-dependent O_2 evolution in isolated thylakoid membranes of *A. nidulans* containing membrane-bound ferredoxin-dependent nitrate and nitrite reductase activities (CANDAU et al. 1976) and of *Anabaena* 7119 supplemented with soluble nitrite reductase (ORTEGA et al. 1976), as well as in *A. nidulans* cells unable to assimilate the resulting ammonium by treatment with L-methionine D,L-sulfoximine (FLORES et al. 1983a). Two mol of O_2 were evolved per mol of NO_3^- reduced to NH_4^+ (8 reducing equivalents), according to the equation:

$$NO_3^- + 4H_2O + 2H^+ \xrightarrow{\text{LIGHT}} NH_4^+ + 2O_2 + 3H_2O$$

Evidence for the direct coupling of nitrate assimilation to water photolysis has also been obtained in *A. nidulans* cells under physiological conditions with simultaneous CO_2 and nitrate assimilation. At saturating CO_2 concentrations, nitrate increases the rate of O_2 evolution in a wide range of light intensities, the stimulation being maximal at saturating light (ROMERO and LARA 1987). As CO_2 fixation is operating at full capacity under these light and CO_2 saturating conditions, the extra O_2 evolution induced by nitrate cannot be due to the refixation of CO_2 produced from carbohydrate oxidation but to photosynthetic nitrate assimilation. It does appear that when CO_2 is the only electron acceptor available, the rate of non-cyclic electron flow is not limited by light but by the rate of CO_2 fixation through the reductive pentose phosphate cycle. Addition of nitrate as a second electron acceptor releases non-cyclic electron flow from this limitation, allowing higher rates of electron transport from water to ferredoxin, the electron donor for nitrate reduction. Nitrate, therefore, increases the maximum capacity of the photosynthetic apparatus (ROMERO and LARA 1987), although the apparent quantum yield of O_2 evolution remains unchanged (ROMERO et al. 1989).

Interestingly, the stoichiometry between the extra O_2 evolved and the NO_3^- taken up under these conditions in *Anacystis* is very low at low light intensity, but it increases with photon flux becoming close to the theoretical value of 2.5 mol of O_2 per mol of NO_3^- assimilated (10 reducing equivalents) at saturating light. This indicates that the stimulation of non-cyclic electron flow induced by nitrate at saturating photon flux density suffices to support the reductant demand for nitrate assimilation, while at low light it does not (ROMERO and LARA 1987).

These features allow to explain the fact that nitrate depresses CO_2 fixation at limiting but not at saturating light, this depression reflecting the competition for photosynthetically generated assimilatory power (ROMERO and LARA 1987, LARA and GUERRERO 1989). This has also been observed in either N-sufficient spinach protoplasts (RATHNAM 1978), or *Scenedesmus* cells (LARSSON et al. 1985) supplied with saturating CO_2 concentrations, in which nitrate assimilation depresses CO_2 fixation at low, but not at high, light intensities. No competition for assimilatory power should, therefore, exist at saturating light, the photosynthetic apparatus being able to generate enough reductant to sustain the simultaneous assimilation of CO_2 and nitrate at full capacity (ROMERO and LARA 1987, LARA and GUERRERO 1989).

It would appear, therefore, that when nitrate is the nitrogen source assimilated by the cells, the actual effect on the rate of CO_2 fixation should result from the balance between the negative effect of nitrate reduction (reflecting competition for assimilatory power) and the positive effect of ammonium incorporation to carbon skeletons, the preponderance of each effect being dependent on the light intensity. This is consistent with the observation that when ammonium assimilation is inhibited by treatment of the cells with L-methionine D,L-sulfoximine, the nitrate-induced depression of CO_2 fixation at limiting light is potentiated (ROMERO and LARA 1987, CORONIL et al. 1993).

4.2 Regulatory role of CO_2 fixation on nitrate transport

Although the assimilation of either ammonium or nitrate exhibits an obligate requirement for CO_2 fixation, the dependence of nitrate assimilation upon the provision of fixed carbon is much more stringent than that exhibited by ammonium assimilation. Thus, in *A.*

nidulans cells, ammonium assimilation reaches its maximum rate at photon flux densities allowing only about one-third of the maximum rate of CO_2 fixation. In contrast, the rates of nitrate assimilation and CO_2 fixation at every light intensity correlate very tightly, indicating that in addition to the carbon skeletons demand for amino acid biosynthesis, an extra carbon requirement exists for nitrate assimilation (LARA and ROMERO 1986). Other observations illustrate the tight correlation between nitrate assimilation and CO_2 fixation in *Anacystis*. Nitrate assimilation exhibits saturation kinetics with respect to the CO_2 concentration available, reaching half-maximum rate at a CO_2 concentration very close to the $K_{m(CO2)}$ value for CO_2-dependent O_2 evolution (FLORES et al. 1983b). Also, the progressive inhibition of CO_2 fixation by increasing concentrations of D,L-glyceraldehyde results in a gradual decrease in the rate of nitrate assimilation (ROMERO et al. 1985).

This strict carbon dependence of nitrate assimilation seems to be related to its feed-back inhibition by some organic nitrogen product(s), which manifests specially after ammonium addition. Studies with different inhibitors have shown that prevention of ammonium assimilation protects nitrate utilization from the negative effects of ammonium and eliminate the requirement for CO_2 fixation (FLORES et al. 1983b, ROMERO et al. 1985). A common link appears, therefore, to underlay the carbon requirement and the ammonium-dependent inhibition of nitrate assimilation, and, in fact, there is evidence indicating that either carbon reserves or high CO_2 fixation rates counteract the negative effects of ammonium addition on nitrate assimilation (FLORES et al. 1983b, ROMERO et al. 1987, LARA and GUERRERO 1989). A model for a concerted feed-back regulation of nitrate assimilation has been proposed involving organic nitrogen compound(s) generated from nitrate via ammonium. Activity of the system will depend on the level of such inhibitory metabolites, determined in turn by the balance between their generation by nitrate assimilation, via ammonium, and their removal by combination with CO_2 fixation products (FLORES et al. 1983b, ROMERO et al. 1985, 1987, GUERRERO and LARA 1987).

The target of such a regulatory mechanism in *A. nidulans* is nitrate transport across the plasma membrane. In this cyanobacterium nitrate is actively transported into the cell, most probably through a nitrate/sodium symport system driven by the energy of the $\Delta\mu Na^+$ (LARA et al. 1987, RODRÍGUEZ et al. 1992, LARA et al. 1993). The nitrate transport activity is sensitive to a rapid and reversible modulation exerted by products of both ammonium and CO_2 assimilation, in such a way that it is effectively inhibited by either ammonium addition or by depressing CO_2 fixation with D,L-glyceraldehyde, and the inhibition in either case is prevented by blocking ammonium assimilation with L-methionine D,L-sulfoximine (LARA et al. 1987). This supports the common basis of the mechanism underlying the antagonistic effects of ammonium and CO_2 assimilation on nitrate transport, and provides evidence that photosynthetic nitrate assimilation in cyanobacteria is primarily controlled at the level of substrate supply to the cell (LARA et al. 1987, GUERRERO et al. 1990). This information provides the basis to understand the carbon dependence of nitrate assimilation in these organisms.

A similar situation appears to be operative in some eukaryotic microalgae, such as the diatom *Phaeodactylum tricornutum*, in which intracellular nitrate accumulation is inhibited by ammonium addition or CO_2 deprivation (CRESSWELL and SYRETT 1979, SYRETT 1988), and the green alga *Ankistrodesmus braunii*, in which nitrate uptake, but not nitrate reduction, is hampered in the absence of carbon assimilation (ULLRICH and EISELE 1978).

5 Carbon dependence of N_2 fixation

Operativity of the N_2 fixation process in cyanobacteria is tightly linked to carbon metabolism. Aside from the requirement of organic carbon compounds for assimilation of the ammonium resulting from N_2 fixation (discussed in section 3), other aspects of the N_2 fixation process itself can involve the participation of carbohydrates to ensure an adequate performance. These are briefly described in this Section.

5.1 Source of assimilatory power for N_2 fixation

Under autotrophic conditions, photosynthesis ultimately supplies the reductant and ATP needed for N_2 fixation in diazotrophic cyanobacteria. The question arises, however, as to whether photosynthesis directly supports N_2 fixation or the link between these processes is indirect, with fixed carbon compounds acting as intermediates.

The latter appears to be clearly the case for heterocystous cyanobacteria, in which nitrogenase is confined to heterocysts (MURRY et al. 1984, ELHAI and WOLK 1990). The process of heterocyst differentiation also leads to a loss of the photosynthetic water-splitting complex, as well as of the key enzymes of the Calvin cycle, namely ribulose bisphosphate carboxylase-oxygenase and phosphoribulokinase (WINKENBACH and WOLK 1973, CODD and STEWART 1977). As a consequence, heterocysts lack the ability to fix CO_2, and to evolve O_2, in the light, and become dependent on the adjacent vegetative cells for provision of organic carbon compounds. The identity of the imported carbon compounds has not yet been unambiguously assessed, although disaccharides might be the main source of reduced carbon flowing from the vegetative cells to heterocysts (HAURY and SPILLER 1981, JÜTTNER 1983, FAY 1992, BUIKEMA and HASELKORN 1993). Oxidation of these compounds in the heterocysts would provide reducing equivalents for nitrogenase under autotrophic conditions, ATP being most probably supplied by cyclic photophosphorylation (BOTTOMLEY and STEWART 1977), which can take place at the heterocysts.

Aerobic respiration of photosynthetically fixed carbon has also been proposed to be the process supplying the reductant for N_2 fixation in some non-heterocystous cyanobacteria. In batch aerobic cultures of *Gloeothece* incubated under a 12 h-light-12 h-dark regime, nitrogen fixation took place primarily during the dark period, and it was apparently supported by the oxidative breakdown of reserve carbohydrate accumulated during the light phase (GALLON et al. 1988). This and other studies have led to the proposal that the provision of assimilatory power for N_2 fixation in *Gloeothece* is only indirectly linked to photosynthesis (GALLON et al. 1988, GALLON and

CHAPLIN 1988). Nevertheless, continuous cultures of *Gloeothece* fix N_2 predominantly during the light period (ORTEGA-CALVO and STAL 1991), as also do batch cultures of *Gloeothece* incubated microaerobically (GALLON and STAL 1992). It cannot, therefore, be excluded that in *Gloeothece* photosynthesis directly supports N_2 fixation. The latter appears actually to be the case for the marine filamentous non-heterocystous *Trichodesmium,* which both under natural and laboratory conditions exhibits light-dependent N_2 fixation (GALLON and STAL 1992, OHKI et al. 1992).

5.2 Carbohydrates and the protection of nitrogenase from oxygen

The extreme sensitivity of the nitrogenase enzyme system to oxygen, a general problem to all diazotrophs, reaches its maximal expression in the cyanobacteria, with the coexistence in the same organism of two conflicting processes, namely oxygen-evolving photosynthesis and oxygen-sensitive N_2 fixation. Like other diazotrophs, cyanobacteria have developed a variety of mechanisms in order to limit the deleterious effects of oxygen on N_2 fixation (FAY 1992, BUIKEMA and HASELKORN 1993). Carbon-supported respiratory oxygen consumption seems to be one of the most general strategies used by cyanobacteria, both heterocystous and non-heterocystous strains, to reduce the oxygen tension at the site of N_2 fixation (FAY 1992), thus contributing decisively to the operativity of the latter process. A mutant of *Anabaena* PCC 7120 with defective respiration expresses active nitrogenase under anaerobic conditions only, thus indicating the requirement of heterocyst respiration for protection of nitrogenase from inactivation by oxygen (ERNST et al. 1992).

5.3 Carbohydrates and regulation of the nitrogenase level

Nitrogenase activity in heterocystous cyanobacteria is strongly influenced by the size of the reserve carbohydrate pool. Although this correlation can be explained, at least in part, on the basis of a supply of reductant from carbohydrates to N_2 fixation, the carbohydrate pool has also a marked influence on the level of the active nitrogenase enzyme complex.

The sensitivity of nitrogenase to the negative effects of oxygen and of combined nitrogen (ERNST et al. 1984, 1990, GUERRERO and LARA 1987, FAY 1992) varies according to the C/N ratio of the cells. In general, N-starved (carbohydrate-rich) heterocystous cyanobacteria do not exhibit nitrogenase sensitivity to ammonium, but the inhibitory effects of ammonium reappear upon recovery from nitrogen starvation (RAMOS et al. 1985). Cells from different stages of growth (MURRY and BENEMANN 1979, MURRY et al. 1983) or exposed to different light intensities (OHMORI and HATTORI 1974) also shows a different behavior with regard to the responsiveness of nitrogenase to products of ammonium assimilation. A common factor effected by these conditions could be the availability of carbon skeletons. It has been proposed that a high supply of carbohydrates could counteract the negative effects of ammonium on nitrogenase synthesis and activity (RAMOS et al. 1985, GUERRERO and LARA 1987), in a similar fashion as that proposed for the regulation of nitrate assimilation (see section 4).

In addition to being under transcriptional control by cell differentiation, oxygen and combined nitrogen (ELHAI and WOLK 1990, FAY 1992, BUIKEMA and HASELKORN 1993), nitrogenase activity is also negatively influenced by limitation of carbon supply, as well as by oxygen and ammonium (ERNST et al. 1990, CAPONE et al. 1990). Inactivation of the enzyme is correlated with a decrease of electrophoretic mobility of the dinitrogenase reductase (Fe-protein) component, indicative of a posttranscriptional enzyme modification, both in the heterocystous *Anabaena* and in the non-heterocystous *Trichodesmium* (REICH and BÖGER 1989, BÖHM et al. 1992, ZEHR et al. 1993). The function of this modification is not known. It might render nitrogenase more oxygen-tolerant or serve as a switch to give priority to processes other than N_2 fixation under conditions of reductant limitation (ERNST et al. 1992). In *Trichodesmium,* which fixes N_2 under illumination only, the change in molecular mass of the Fe-protein is correlated with the diel change in nitrogenase activity (CAPONE et al. 1990). The trigger process of nitrogenase inactivation in *Anabaena* apparently depends on variations of the glycogen pool (ERNST et al. 1984), and it has been proposed to develop as a result of decreased carbohydrate-supported reductant supply caused by either carbon limitation or by increased diversion of carbohydrates towards ammonium assimilation (ERNST et al. 1990).

5.4 Carbon/Nitrogen ratio and heterocyst differentiation

In heterocystous cyanobacteria, the acquisition of the capacity of fix N_2 is accompanied by the differentiation of vegetative cells into heterocysts. This process involves a whole set of profound structural and biochemical changes (HASELKORN 1978, WOLK 1982, FAY 1992, BUIKEMA and HASELKORN 1993). Heterocyst differentiation and concomitant nitrogenase synthesis are generally suppressed when combined nitrogen is available, and are initiated following a period of nitrogen deficiency, with depletion of cellular nitrogen reserves. As a consequence of nitrogen starvation, the cellular C/N ration increases. KULASOORIYA et al. (1972) studied the evolution of the C/N ratio in filaments of *Anabaena cylindrica* transferred from a medium containing ammonium into a medium free of combined nitrogen, and also followed the time-course of heterocyst differentiation and development of nitrogenase activity. They observed that heterocyst development did not begin until the C/N ratio increased from 4.5 to 6, and that nitrogenase activity was negligible until the ratio rose to about 8, in coincidence with the appearance of mature heterocysts. This and other evidence indicate that the C/N balance is a decisive factor in the control of heterocyst differentiation and subsequent nitrogenase expression, although the underlying mechanisms remain unknown (FAY 1992).

6 Concluding remarks

The assimilation of carbon and nitrogen are closely intertwined, both at the metabolic and the regulatory level. Mutual interactions exist that determine a precise balance of these two main bioelements in the cell. The understanding of how this balance is achieved in the case of cyanobacteria is far to be completed, although a substantial progress has been done during the last few years in the knowledge of some of the interactions involved.

It is by now well established that assimilation of inorganic nitrogen, either as nitrate, ammonium or N_2, induces a diversion of carbon flow to the biosynthesis of amino acids and other nitrogenous compounds, with a concomitant decrease in the carbon flow to glycogen. A stimulation of CO_2 fixation takes place additionally when the source of nitrogen is ammonium. With regard to effects of carbon assimilation on nitrogen metabolism other than the metabolic requirement for carbon skeletons to incorporate ammonium, the dependence of N_2 fixation upon a degradative metabolism of carbohydrates for both provision of reducing power and protection of nitrogenase against oxygen with cellular carbon playing also a relevant role in the determination of nitrogenase level is evident. In contrast to N_2 fixation, nitrate assimilation in cyanobacteria does not require carbohydrate-derived reductant, since nitrate reduction is in these organisms a genuine photosynthetic process that uses photogenerated reduced ferredoxin. Nevertheless, a strict regulatory requirement for active CO_2 fixation is evident in nitrate assimilation. The modulation of the latter process by CO_2 fixation has definitively been established to occur at the level of nitrate transport into the cell.

Despite the significant advances achieved in the understanding of the interactions between the metabolisms of carbon and nitrogen, little is still known about the underlying mechanisms involved, and this awaits further research.

Acknowledgment

Research from the author's laboratory has been financed by the Spanish Directión General de Investigación Científica y Tecnica (grants PB88-0019 and PB91-0611) and Plan Andaluz de Investigación (gr. 3101).

References

Bassham JA, Larsen PO, Lawyer AL, Cornwell KL (1981) Relationship between nitrogen metabolism and photosynthesis. In: Bewley JD (ed) Nitrogen and carbon metabolism. Nijhoff/Junk, The Hague, pp 135-163

Böhm I, Halbherr A, Smaglinski S, Ernst A, Böger P (1992) *In vitro* activation of dinitrogenase reductase from the cyanobacterium *Anabaena variabilis* (ATCC 29413) J Bacteriol 174:6179-6183

Bottomley PJ, Stewart WDP (1977) ATP and nitrogenase activity in nitrogen-fixing heterocystous blue-green algae. New Phytol 79:625-638

Bottomley PJ, Grillo JF, van Baalen C, Tabita FR (1979) Synthesis of nitrogenase and heterocysts by *Anabaena* sp. CA in the presence of high levels of ammonia. J Bacteriol 140:938-943

Boussiba S, Resch CM, Gibson J (1984) Ammonia uptake and retention in some cyanobacteria. Arch Microbiol 138:287-292

Buikema WJ, Haselkorn R (1993) Molecular genetics of cyanobacteria development. Annu Rev Plant Physiol Plant Mol Biol 44:33-52

Candau P, Manzano C, Losada M (1976) Bioconversion of light energy into chemical energy through reduction with water of nitrate to ammonia. Nature 262:715-717

Capone DG, O'Neil JM, Zehr J, Carpenter EJ (1990) Basis for diel variation in nitrogenase activity in the marine planktonic cyanobacterium *Trichodesmium thiebautii*. Appl Environ Microbiol 56:3532-3536

Codd GA, Stewart WDP (1977) Ribulose-1,5-bisphosphate carboxylase in heterocysts and vegetative cells of *Anabaena cylindrica*. FEMS Microbiol Lett 2:247-249

Coronil T, Lara C, Guerrero MG (1993) Shift in carbon flow and stimulation of amino-acid turnover induced by nitrate and ammonium assimilation in *Anacystis nidulans*. Planta 189:461-467

Cresswell RC, Syrett PJ (1979) Ammonium inhibition of nitrate uptake by the diatom *Phaeodactylum tricornutum*. Plant Sci Lett 14:321-325

Elhai J, Wolk CP (1990) Developmental regulation and special pattern of expression of the structural genes for nitrogenase in the cyanobacterium *Anabaena*. EMBO J 9:3379-3388

Elrifi IR, Turpin DH (1987) The path of carbon flow during NO_3^- induced photosynthetic suppression in N-limited *Selenastrum minutum*. Plant Physiol 83:97-104

Ernst A, Kirschenlohr H, Díez J, Böger P (1984) Glycogen content and nitrogenase activity in *Anabaena variabilis*. Arch Microbiol 140: 120-125

Ernst A, Reich S, Böger P (1990) Modification of dinitrogenase reductase in the cyanobacterium *Anabaena variabilis* due to C starvation and ammonia. J Bacteriol 172:748-755

Ernst A, Black T, Cai Y, Panoff J-M, Tiwari DN, Wolk CP (1992) Synthesis of nitrogenase in mutants of the cyanobacterium *Anabaena* sp. strain PCC 7120 affected in heterocyst development or metabolism. J Bacteriol 174:6025-6032

Fay P (1992) Oxygen relations of nitrogen fixation in cyanobacteria. Microbiol Rev 56:340-373

Flores E, Guerrero ML, Losada M (1983a) Photosynthetic nature of nitrate uptake and reduction in the cyanobacterium *Anacystis nidulans*. Biochim Biophys Acta 722:408-416

Flores E, Romero JM, Guerrero MG, Losada M (1983b) Regulatory interaction of photosynthetic nitrate utilization and carbon dioxide fixation in the cyanobacterium *Anacystis nidulans*. Biochim Biophys Acta 725:529-532

Fogg GE, Stewart WDP, Fay P, Walsby AE (eds) (1973) The Blue-green Algae. Academic Press, London, pp 190-213

Gallon JR, Perry SM, Rajab TMA, Flayeh KAM, June JS, Chaplin AE (1988) Metabolic changes associated with the diurnal pattern of N_2 fixation in *Gloeothece*. J Gen Microbiol 134:3079-3087

Gallon JR, Chaplin AE (1988) Nitrogen fixation. In: Rogers LJ, Gallon JR (eds) Biochemistry of the Algae and Cyanobacteria. Clarendon Press, Oxford, pp 147-173

Gallon JR, Stal LJ (1992) N_2 fixation in non-heterocystous cyanobacteria. In: Carpenter EJ, Capone DG, Rueter JG (eds) Marine Pelagic Cyanobacteria: *Trichodesmium* and other Diazotrophs. Kluwer, Dordrecht, pp 115-139

García-González M, Sivak MN, Guerrero MG, Preiss J, Lara C (1992) Depression of carbon flow to the glycogen pool induced by nitrogen assimilation in intact cell of *Anacystis nidulans*. Physiol Plant 86:360-364

Guerrero MG, Lara C (1987) Assimilation of inorganic nitrogen. In: Fay P, van Baalen C (eds) The Cyanobacteria. Elsevier Science Publishers, Amsterdam, pp 163-186

Guerrero MG, Romero JM, Rodríguez R, Lara C (1990) Nitrate transport in cyanobacteria. In: Ullrich WR, Rigano C, Fuggi A, Aparicio PJ (eds) Inorganic Nitrogen in Plants and Microorganisms. Springer, Berlin, pp 79-85.

Haselkorn R (1978) Heterocysts. Annu Rev Plant Physiol 29:319-344

Haury JF, Spiller H (1981) Fructose uptake and influence on growth of and nitrogen fixation by *Anabaena variabilis*. J Bacteriol 147:227-235

Jüttner J (1983) ^{14}C-labelled metabolites in heterocysts and vegetative cells of *Anabaena cylindrica* filaments and their presumptive function as transport vehicles of organic carbon and nitrogen. J Bacteriol 155:628-633

Kanazawa TK, Kanazawa MR, Kirk MR, Bassham JA (1972) Regulatory effects of ammonia on carbon metabolism in *Chlorella pyrenoidosa* during photosynthesis and respiration. Biochim Biophys. Acta 265:656-669

Kulasooriya SA, Lang NJ, Fay P (1972) The heterocysts of blue-green algae III. Differentiation and nitrogenase activity. Proc R Soc London Ser B 181:199-209

Lara C (1992) Photosynthetic nitrate assimilation interactions with CO_2 fixation. In: Barber J, Medrano H, Guerrero MG (eds) Trends in photosynthesis Research, Intercept, London, pp 195-208

Lara C, Guerrero MG (1989) The photosynthetic assimilation of nitrate and its interactions with CO_2 fixation. In: Barber J, Malkin R (eds) Techniques and new developments in Photosynthesis Research. Plenum, New York, pp 393-411

Lara C, Romero JM (1986) Distinctive light and CO_2 fixation requirements of nitrate and ammonium utilization by the cyanobacterium *Anacystis nidulans*. Plant Physiol 81:686-688

Lara C, Romero JM, Guerrero MG (1987) Regulated nitrate transport in the cyanobacterium *Anacystis nidulans*. J Bacteriol 169:4376-4378

Lara C, Rodríguez R, Guerrero MG (1993) Sodium-dependent nitrate transport and energetics of cyanobacteria. J Phycol 29:389-395

Larsson M, Olsson T, Larsson C-M (1985) Distribution of reducing power between photosynthetic carbon and nitrogen assimilation in *Scenedesmus*. Planta 164:246-253

Lawrie AC, Codd GA, Stewart WDP (1976) The incorporation of nitrogen into products of recent photosynthesis in *Anabaena cylindrica*. Lemm. Arch Microbiol 107:15-24

Levy C, Preiss J (1976) Regulatory properties of the ADP-glucose pyrophosphorylase of the blue-green bacterium *Synechococcus* 6301. Plant Physiol 58:753-756

Miller AG, Espie GS, Canvin DT (1990) Physiological aspects of CO_2 and HCO_3^- transport in cyanobacteria: a review. Can J Bot 68:1291-1302

Murry MA, Benemann JR (1979) Nitrogenase regulation in *Anabaena cylindrica*. Plant Cell Physiol 20:1391-1401

Murry MA, Jensen DB, Benemann JR (1983) Role of ammonia in regulation of nitrogenase synthesis and activity in *Anabaena cylindrica*. Biochim Biophys Acta 756:13-19

Murry MA, Hallenbeck PC, Benemann JR (1984) Immunochemical evidence that nitrogenase is restricted to heterocysts in *Anabaena cylindrica*. Arch Microbiol 137:194-199

Ohki K, Zehr JP, Fujita Y (1992) *Trichodesmium*: Establishment of culture and characteristics of N_2 fixation. In: Carpenter EJ, Capone DG, Rueter JG (eds) Marine pelagic cyanobacteria: *Trichodesmium* and other diazotrophs. Kluwer, Dordrecht, pp 307-318

Ohmori M, Hattori A (1972) Effect of nitrate on nitrogen-fixation by the blue-green alga *Anabaena cylindrica*. Plant Cell Physiol 13:589-599

Ohmori M, Hattori A (1974) Effect of ammonia on nitrogen-fixation by the blue-green alga *Anabaena cylindrica*. Plant Cell Physiol 15:131-142

Ortega T, Castillo F, Cardenás J (1976) Photolysis of water coupled to nitrate reduction by *Nostoc muscorum* subcellular particles. Biochem Biophys Res Commun 71:885-891

Ortega-Calvo JJ, Stal LJ (1991) Diazotrophic growth of the unicellular cyanobacterium *Gloeothece* sp. PCC 6909 in continuous culture. J Gen Microbiol 137:1789-1797

Paul JS, Cornwell KL, Bassham JA (1978) Effects of ammonia on carbon metabolism in photosynthesizing isolated cells from *Papaver somniferum* L. Planta 142:49-54

Ramos JL, Madueño F, Guerrero MG (1985) Regulation of nitrogenase levels in *Anabaena* sp. ATCC 33047 and other filamentous cyanobacteria. Arch Microbiol 141:105-111

Rathnam CKM (1978) Malate and dihydroxyacetone phosphate-dependent nitrate reduction in spinach leaf protoplasts. Plant Physiol 62:220-223

Reich S. Böger P (1989) Regulation of nitrogenase activity in *Anabaena variabilis* by modification of the Fe-protein. FEMS Microbiol Lett 58:81-86

Rodríguez R, Lara C, Guerrero MG (1992) Nitrate transport in the cyanobacterium *Anacystis nidulans* R2. Kinetic and energetic aspects. Biochem J 282:639-643

Romero JM, Lara C (1987) Photosynthetic assimilation of NO_3^- by intact cells of the cyanobacterium *Anacystis nidulans*. Influence of NO_3^- and NH_4^+ assimilation on CO_2 fixation. Plant Physiol 82:208-212

Romero JM, Lara C, Guerrero MG (1985) Dependence of nitrate utilization upon active CO_2 fixation in *Anacystis nidulans*: A regulatory aspect of the interaction between photosynthetic carbon and nitrogen metabolism. Arch Biochem Biophys 237:396-401

Romero JM, Coronil T, Lara C, Guerrero MG (1987) Modulation of nitrate uptake in *Anacystis nidulans* by the balance between ammonium assimilation and CO_2 fixation. Arch Biochem Biophys 256:578-584

Romero JM, Lara C, Sivak MN (1989) Changes in net O_2 exchange induced by inorganic nitrogen in the blue-green alga *Anacystis nidulans*. Plant Physiol 91:28-30

Smith RG, Vanlerberghe GC, Stitt M, Turpin DH (1989) Short-term metabolite changes during transient ammonium assimilation by the N-limited green alga *Selenastrum minutum*. Plant Physiol 91:749-755

Stanier RY, Cohen-Bazire G (1977) Photosynthetic prokaryotes: the cyanobacteria. Annu Rev Microbiol 31:225-274

Syrett PJ (1988) Uptake and utilization of nitrogen compounds. In: Rogers LJ, Gallon JR (eds) Biochemistry of the Algae and Cyanobacteria. Clarendon Press, Oxford, pp 23-29

Ullrich WR, Eisele R (1978) Relations between nitrate uptake and nitrate reduction in *Ankistrodesmus braunii*. In: Ducet G. Heller R. Tellier M (eds) Transmembrane ionic exchanges in plants. CNRS, Paris, pp 307-315

Walker DA, Sivak MN (1986) Photosynthesis and phosphate: A cellular affair? Trends Biochem Sci 11:176-179

Winkenbach F, Wolk CP (1973) Activities of enzymes of the oxidative and reductive pentose phosphate pathways in heterocysts of a blue-green alga. Plant Physiol 52:480-483.

Wolk CP (1982) Heterocysts. In : Carr NG, Whitton BA (eds) The Biology of cyanobacteria. Blackwell, Oxford, pp 359-386

Woo KC, Canvin DT (1980a) Effect of ammonia on photosynthetic carbon fixation in isolated spinach leaves cells. Can J Bot 58:505-510.

Woo KC, Canvin DT (1980b) Effect of ammonia, nitrite, glutamate, and inhibitors of N metabolism on photosynthetic carbon fixation in isolated spinach leaves cells. Can J Bot 58:511-516

Zehr JP, Wyman M, Miller V, Duguay L, Capone DG (1993) Modification of the Fe protein of nitrogenase in natural populations of *Trichodesmium thiebautii*. Appl Environ Microbiol 59:669-676

Environmental Stresses

MARTIN HAGEMANN and NORBERT ERDMANN

University of Rostock, Department of Biology, Doberaner Str 143, D-18051, Germany

1 Introduction

Like all living cells, cyanobacteria are open systems, interacting with their environment by the constant exchange of matter, energy and information. For each environmental factor, there is a range of levels (the tolerance range) within which the cells are capable of growth, the rate of growth following an optimum curve as the level of that factor changes. However, cells are capable of survival over a larger range of variation, the resistance range. Stress is defined as a detrimental change in an organism caused by environmental factors. The factor inducing the change is called a stress factor or stressor.

The response to a stressor is a physiological adaptation (acclimation) and, depending on the kind and intensity of the factor, can persist for hours or days in cyanobacteria (Fig. 1). It starts with destabilization of the cellular metabolism due, say, to loss of membrane semipermeability, denaturation of proteins, changes in metabolic balance, formation of radicals, etc., leading to growth inhibition. This is followed by an acclimation phase initiated by a

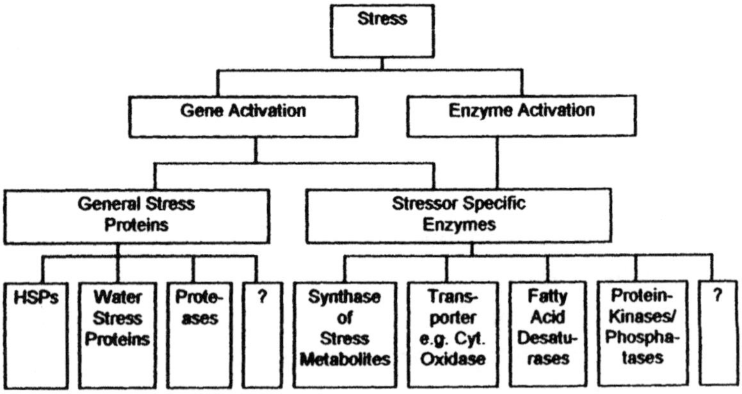

Fig. 1. Diagram illustrating the basic processes involved in cyanobacterial adaptation strategies

change in gene expression or non-genetic mechanisms which trigger the synthesis of low-molecular stress metabolites, modify the membrane composition, induce certain transport processes or lead to appropriate metabolic changes, etc. (Fig. 2). At the same time, changes in gene expression lead to the synthesis of various stress proteins. Some of these, having a general protective function and being concomitants of several stress syndromes, can be considered general stress proteins, but others are stressor-specific.

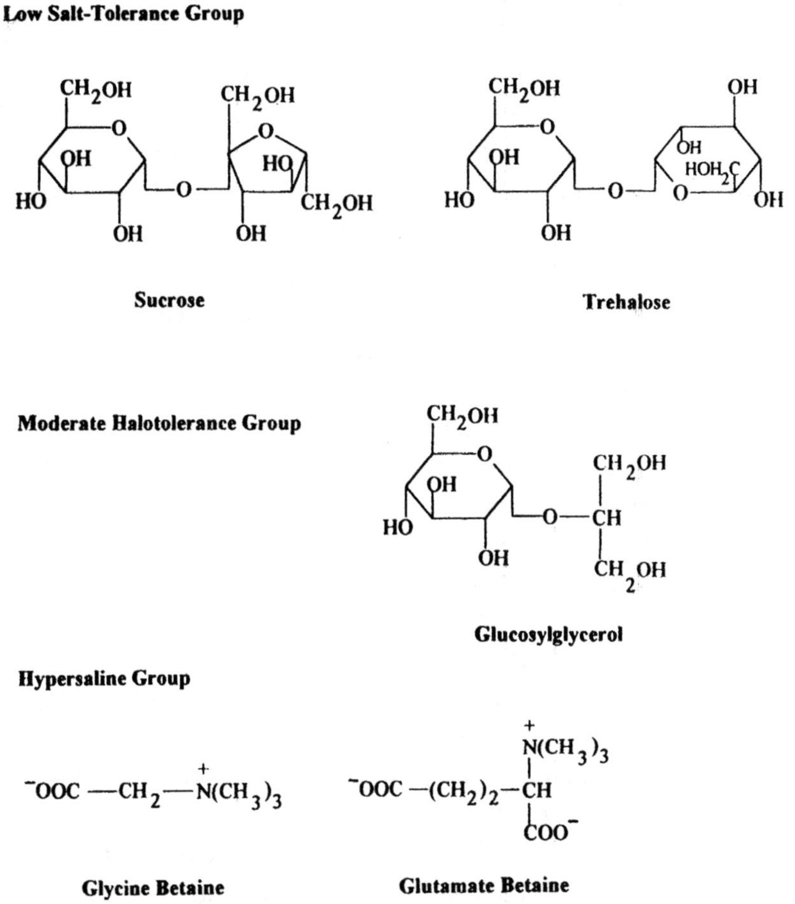

Fig. 2. Formulas of osmoprotective compounds used by cyanobacteria

All of these processes culminate in a recovery phase in which growth is gradually resumed until, finally, a new steady state is achieved. The growth rate of fully acclimated cells is often lower than the initial growth rate, but it can also be higher (acclimation to suboptimal or more optimal environmental conditions). If the intensity of a stressor exceeds the resistance limits of the cell, certain functions break down and death may ensue. However, successful acclimation sometimes confers increased resistance to inhospitable conditions.

2 Individual stress syndromes in cyanobacteria

2.1 General acclimation ability

Cyanobacteria have successfully colonized a wide range of biotopes in the course of their evolution. Today, they are found in virtually all aquatic ecosystems (from freshwater through oceanic to hypersaline, hot springs and ice), in the soil, on naked rock, in deserts and even in the air (FOGG et al. 1973). They owe their worldwide distribution to the wide range of morphological and physiological properties they have acquired and their remarkable capacity to adapt to changes in a wide variety of environmental factors (STANIER and COHEN BAZIRE 1977).

In their natural environment, cyanobacteria generally exist under growth-limiting conditions. Several environmental parameters such as light, salinity, nutrient availability and temperature change frequently. Few data are available regarding the adaptation of cyanobacteria in natural populations. Most studies of their adaptation to the various environmental parameters have been performed with selected model strains in laboratory cultures under near-optimum growth conditions. In the past few years, many of the adaptation mechanisms used by cyanobacteria have been described down to the molecular level, including changes in response to varying light quality (chromatic adaptation) and nitrogen deficiency (heterocyst differentiation) (TANDEAU De MARSAC 1991, BUIKEMA and HASELKORN 1993, TANDEAU De MARSAC and HOUMARD 1993).

This chapter will deal with physiological adaptation processes (non-genetic adaptation, acclimation-cf. REED 1984) of cyanobacteria to the natural environmental factors salt, temperature and light. Little research has been done so far on the direct effects of different stress factors on the processes involved in N metabolism. However, it can safely be assumed that the metabolic changes described in the context of general strategies for adapting to various environmental factors will be reflected at least indirectly in nitrogen metabolism.

2.2 Salt, osmotic and desiccation stress

2.2.1 General remarks

Salinity is one of the most important abiotic factors in aquatic biotopes. An increase in salinity (a reduction has analogous, but opposite effects) represents a combination of two stress situations for cyanobacteria. In the first place, the reduction in the water potential of the surrounding medium causes the cells to lose water. Osmotic and drought stress also have this effect, and the adaptation mechanisms of cyanobacteria for coping with all three stressors therefore share many features in common. Water losses lead to shrinkage of cells without cell walls and loss of turgor in those with them. In contrast to pure osmotic stress (caused by non-permeable organic osmotic agents) and desiccation, an increase in salinity means a dramatic increase in concentrations of inorganic ions (especially Na^+ and Cl^-) in the surrounding medium. These ions enter the cell along chemical and, in some cases, electrochemical gradients, thus counteracting the loss of water since this process reduces the osmotic potential difference. However, high concentrations of inorganic ions are toxic, interfering with cell metabolism except in halobacteria (KUSCHNER 1985). During adaptation to salt stress, a balanced osmotic potential is achieved by the extrusion of excess inorganic ions from the cell and the accumulation of so called osmoprotective substances.

2.2.2 Accumulation of osmoprotective substances

Osmoprotective substances are low-molecular hydrophilic compounds that accumulate in large quantities in salt-loaded cells. They do not interfere with the cell metabolism because they are largely inert and carry no net charge (BROWN 1976). They reduce the osmotic potential of the cytoplasma to prevent further loss of water and apparently prevent denaturation by helping macromolecules to retain their natural configurations. Chemically, they can be grouped into non-electrolytes and electrolytes (Fig. 3). The non-electrolytes consist of polyols and carbohydrates and are common in algae, yeasts and cyanobacteria. Except for 3-dimethylsulfoniopropionate and dimethylthetine, the electrolytes are quaternary compounds of ammonium and are typical of plants, bacteria and cyanobacteria (CSONKA and HANSON 1991).

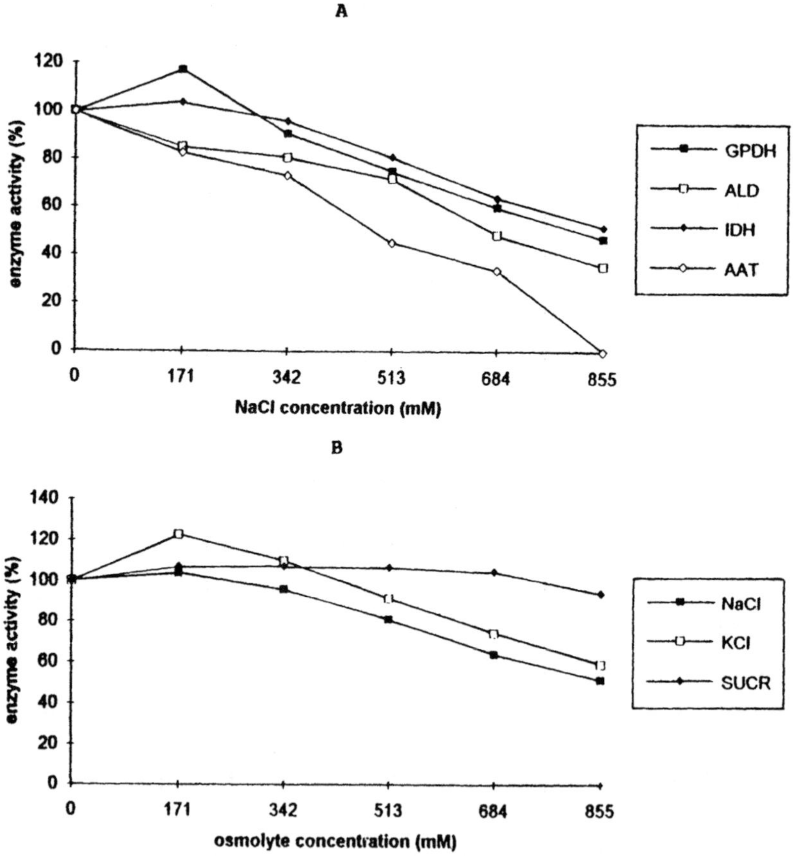

Fig. 3. Inhibitory action of increasing salt concentrations and compatibility of increasing sucrose concentrations *in vitro* on enzyme activities in *Microcystis firma* (HAGEMANN, unpubl., cf. HAGEMANN et al. 1989a for methods). **A** Increasing NaCl concentrations *in vitro* (IDH- isocitrate dehydrogenase, GDPH- glucose 6-phosphate dehydrogenase, ALD- fructose bisphosphate aldolase, AAT- aspartate aminotransferase); **B** Comparison of the effects of increasing NaCl, KCl and sucrose concentrations *in vitro* on isocitrate dehydrogenase

In 1980, the first osmoprotective substance of cyanobacteria was found in salt-loaded cells of the strain *Synechococcus* NM100 by means of the ^{13}C-NMR technique. The substance was the heteroside glucosylglycerol discovered in cyanobacteria only a year earlier (KOLLMAN 1979), and its concentration increased in linear proportion to the salinity (BOROWITZKA et al. 1980). Two disaccharides, sucrose (BLUMWALD and TEL-OR 1982, ERDMANN 1983) and trehalose (REED and STEWART 1983) were identified as further cyanobacterial osmoprotective substances in the subsequent years.

Nitrogen-containing osmoprotective substances were of special interest since these are associated with the highest tolerances and are widespread among prokaryotes and eukaryotes. So far, two quaternary ammonium compounds have been found in cyanobacteria: glycine betaine (MOHAMMAD et al. 1983) and glutamate betaine (MACKAY et al. 1984). The accumulation of glycine betaine in cyanobacteria in proportion to the intensity of a stressor has been reported by various authors (REED et al. 1984a, GABBAY-AZARIA et al. 1988), the highest accumulation rate observed so far being 60 nmol (mg protein)$^{-1}$h^{-1}; this is one order of magnitude higher than in plants (ISHITANI et al. 1993).

In the meantime, about 130 cyanobacterial strains have been studied for the synthesis of osmoprotective substances in response to salt loading. Sucrose has been described as the main osmolyte in 66 strains, trehalose in 20, glucosylglycerol in 24 and glycine or glutamate betaine in 18 (ERDMANN 1983, MACKAY et al. 1984, REED et al. 1984a, b, REED and STEWART 1985, STAL and REED 1987).

Although a large number of cyanobacterial strains has been studied, no link has been found between the kind of osmoprotective substance accumulated and either taxonomic group or the biotope of origin of the strains. The genus *Synechococcus*, for instance, contains strains capable of producing all osmoprotective substances hitherto described for cyanobacteria, whereas only sucrose-accumulating strains have been found in the genera *Anabaena* and *Nostoc* (REED et al. 1986a). However, it has been possible to assign the strains to three salt resistance groups which differ distinctly in the osmoprotective substances they accumulate (REED et al. 1986a, Fig. 2):

1. The least halotolerant strains accumulate sucrose and trehalose and can tolerate up to 0.7 M NaCl.
2. Moderately halotolerant cyanobacteria accumulate glucosyl-glycerol, and their tolerance limit is 1.8 M NaCl.
3. The highest halotolerance is exhibited by strains that accumulate mainly glycine and glutamate betaine; these strains can tolerate salt concentrations up to 2.7 M.

In many strains, traces of other osmoprotective substances have been detected at the same time as the main osmoprotective substance. For instance, almost all strains accumulating predominantly betaines also exhibit low concentrations of glucosylglycerol, sucrose and/or trehalose. Some strains synthesize sucrose and trehalose simultaneously, and several strains have been found containing sucrose and/or trehalose besides glucosylglycerol (MACKAY et al. 1984, REED et al. 1984a, b). In laboratory cultures of *Synechocystis* sp. PCC 6714 (WARR et al. 1985b), the small amount of sucrose normally found besides glucosylglycerol among the osmoprotective substances increased to 28% at relatively low salinities and elevated temperatures. Although most studies have been performed with laboratory cultures, glucosylglycerol, sucrose and trehalose have also been isolated from cyanobacterial mats taken straight from the North Sea (STAL and REED 1987).

Desiccation also leads to the accumulation of trehalose and sucrose as osmoprotective substances to counteract water losses in drought-tolerant cyanobacteria (HILL et al. 1994). The drought resistance of *Phormidium autumnale* and a *Croococcidiopsis* species on the one hand and the lack of it in *Plectonema boryanum* has been attributed to the ability and lack of such an ability, respectively to accumulate trehalose and sucrose (HERSHKOVITZ et al. 1991a).

So far, few studies have investigated the accumulation of osmoprotective substances in cyanobacteria exposed to high concentrations of organic osmolytes (pure osmotic stress). However, *Aphanothece halophytica* has been shown to accumulate glycine betaine when loaded with sorbitol as an organic osmolyte, although the tolerance level under sorbitol stress was lower than under NaCl stress (ISHITANI et al. 1993).

2.2.3 Protective functions of nitrogen-containing osmoprotective substances

It has been shown *in vivo* and *in vitro* on many occasions that osmoprotective substances are compatible, i.e., they protect rather than inhibit enzymes at physiological concentrations (Fig. 3). WARR et al. (1988) tested the compatibility of various cyanobacterial osmoprotective substances and discovered that their compatibility with the model enzyme glutamine synthetase isolated from 17 freshwater, marine and hypersaline isolates decreases in the order glycine betaine, polyol derivatives, disaccharides. Excessive sucrose concentrations inhibit the enzyme (WARR et al. 1984). However, only glycine betaine has been found to have a haloprotective function (WARR et al. 1984, 1988, GABBAY-AZARIA et al. 1988). Glycine betaine has also been found to protect the cyanobacterial enzymes glucose-6- phosphate dehydrogenase (GABBAY-AZARIA et al. 1988) and RuBisCO (INCHAROENSAKDI et al. 1986) from the inhibitory action of Na^+ and K^+. The harmful effects of Cl^- on RuBisCO can also be cancelled by glycine betaine (ISHITANI et al. 1993).

Apart from protecting purified enzymes, glycine betaine also stabilizes O_2 development and ATP synthesis by thylakoids isolated from *Synechocystis* sp. PCC 6803 if added during handling (MAMEDOV et al. 1991) and generally enhances heat tolerance and resistance to high detergent concentrations (PALEG et al. 1985, SANTORO et al. 1992). Direct bonding of glycine betaine to proteins and its accumulation in the protein-water interface have so far been excluded as protective mechanisms (POLLARD and WYN JONES 1979).

It must be pointed out that the results of all these *in vitro* experiments can be applied only as models to intact organisms because in some cyanobacteria glutamine synthetase is not inhibited by the relatively low concentrations of Na^+, K^+ and Cl^- ions in salt-adapted cells (WARR et al. 1984, 1988).

2.2.4 Biosynthesis of osmoprotective substances

Cyanobacteria are able to synthesize osmoprotective substances themselves, whereas eubacteria accumulate glycine betaine and proline from their surroundings (STROM et al. 1986). The main source of carbon for the biosynthesis of osmoprotective substances is carbon fixation by photosynthesis. However, stored carbohydrates (glycogen) can also be used as a carbon source (WARR et al. 1985a, TEL-OR et al. 1986, MACKAY and NORTON 1987, ERDMANN et al. 1989, KERBY et al. 1990). It has been shown by means of the ^{13}C NMR technique that over 90% of the glucosylglycerol synthesized after a salt shock incorporated photosynthetically fixed carbon and that no more than 10% of the carbon stemmed from glycogen immediately after the shock (TEL-OR et al. 1986, MACKAY and NORTON 1987). Glycogen is also implicated in the synthesis of sucrose by *Synechococcus* sp. 6301, especially in the first few hours following salt shock (KERBY et al. 1990). The role of photosynthesis as a prerequisite for the rapid accumulation of osmoprotective substances is doubtless among the main reasons why cyanobacteria can adapt to salt only in the light, but not in darkness (MACKAY and NORTON 1987, ERDMANN et al. 1989). However, salt loaded cyanobacteria can synthesize small amounts of glucosylglycerol from glycogen or exogenic glucose even in the dark (ERDMANN et al. 1989).

Once reached in the adapted cell, the final concentration of osmoprotective compounds, which is proportional to the salt load, remains constant. There is no turnover of the glucosylglycerol pool in *Microcystis firma* (HAGEMANN et al. 1987). In fact, the absence of turnover is a characteristic property of osmoprotective substances: they are not part, but an end product of the cell metabolism.

A system serving the uptake of glycine betaine has been found in strains that accumulate this substance, but not in those accumulating carbohydrates or polyols. Like the mechanisms in eubacteria, the system is able to import exogenous glycine betaine into the cell in osmotically significant amounts. Unlike the heterotrophic bacteria, cyanobacteria are able to synthesize glycine betaine *de novo*, and the transport system therefore probably serves the recovery of glycine betaine that has diffused out of the cell (MOORE et al. 1987).

The osmoprotective substances vanish rapidly after a hypoosmotic shock. Most of the glucosylglycerol, sucrose and/or nitrogen containing osmoprotective compounds is extruded into the medium after a sudden major reduction in salinity (REED et al. 1986b, FULDA et al. 1990, ERDMANN et al. 1992). This obviously prevents the cell from rupturing under the pressure of an osmotically induced inflow of water. Extrusion appears to take place through stretch-activated channels which are even selective to some degree. *Salmonella typhimurium*, for instance, extrudes all betaine under conditions at which 45% of the serine is retained (KOO et al. 1991). However, at least part of the carbon in osmoprotective substances is incorporated into glycogen after a hypoosmotic salt shock (WARR et al. 1985a, TEL- OR et al. 1986). An α-glucosidase activity that might be implicated in the metabolization of glucosylglycerol has been found in hypoosmotically loaded *Spirulina platensis* cells (MARTEL et al. 1992).

Cyanobacteria in a salt-free medium contain only traces of osmoprotective substances. The synthesis of these compounds commences without a lag-phase after a salt shock (REED et al. 1985b, HAGEMANN et al. 1987), and cell growth is greatly inhibited while they are being accumulated (TEL-OR et al. 1986, ERDMANN et al. 1992). The activation of the GG synthesis mechanism can not be inhibited with chloramphenicol (HAGEMANN et al. 1990). This suggests that the enzyme systems involved are dormant during salt-free conditions and are activated biochemically after a salt shock. The pathway by which GG is synthesized was recently described (HAGEMANN and ERDMANN 1994):

ADP-glucose + glycerol-3-phosphate —(1) → glucosylglycerol-phosphate —(2) → glucosylglycerol

(1) Glucosylglycerol-phosphate synthase

(2) Glucosylglycerol-phosphate phosphatase

In contrast to the synthesis of the structurally related isofloridoside in *Poterioochromonas* (KAUSS 1987), glucosylglycerol synthesis is strictly ADP-glucose dependent, but independent of UDP-glucose. The enzymes involved can be activated by increasing the salinity both *in vitro* and *in vivo*. This process is reversible, i.e. the enzyme can be inactivated by reducing the salinity and then reactivated again by increasing it (HAGEMANN and ERDMANN 1994). The biochemical activation mechanism is not yet known, but the reversibility of the process indicates that the mechanism is not the same as in *Poterioochromonas,* in which the enzyme is activated by partial proteolysis (BRUNNER and KAUSS 1988). Increasing salinity has also been found to activate other cyanobacterial enzymes involved in salt adaptation *in vitro* (GABBAY-AZARIA et al. 1992, 1994).

The pathways by which trehalose, sucrose and glycine betaine are synthesized by cyanobacteria have not yet been studied in detail. However, *Escherichia coli* is known to synthesize trehalose as an osmoprotective substance from UDP glucose and glucose-1-phosphate, and the enzymes involved can also be activated *in vitro* by salt loading (GIAEVER et al. 1988). Many higher plants also produce sucrose as an osmoprotective substance. In this case, the principal precursors are ADP-glucose and fructose-6-phosphate, and the activity of the sucrose phosphate synthase is regulated by reversible changes in its phosphorylation status (HUBER and HUBER 1992).

Higher plants that accumulate glycine betaine synthesize this substance mainly in the chloroplasts. The pathway they use involves the methylation of phosphoryl ethanolamine to phosphoryl choline, dephosphorylation to choline and the gradual oxidation of the choline through betaine aldehyde to glycine betaine (HANSON et al. 1985). The oxidation of choline by betaine aldehyde dehydrogenase is also a necessary step in the biosynthesis of glycine betaine in *Spirulina subsalsa* (GABBAY-AZARIA and TEL-OR 1993), but the corresponding enzyme used by cyanobacteria has still to be characterized. SIBLEY and YOPP (1987) have studied S-adenosyl homocysteine hydrolase in the cyanobacterium *Aphanothece halophytica* and suggest that it is implicated in the regulation of glycine betaine synthesis. This enzyme catalyzes the reversible conversion of S-adenosyl cysteine, which is implicated in glycine betaine biosynthesis as a competitive inhibitor of trans-methylation. In salt-shocked cells, i.e. when demand for glycine betaine suddenly increases, the decomposition of S-adenosyl cysteine is stimulated by the inflow of K^+ ions into the cell, and this might activate the transmethylation processes needed for the biosynthesis of glycine betaine.

2.2.5 Changes in concentrations of inorganic ions

Besides reducing the osmotic potential, salt stress is accompanied by a dramatic increase in the concentrations of inorganic ions (especially Na^+ and Cl^-) in the surrounding medium. Although very high concentrations of inorganic ions are toxic to cyanobacteria, slight Na^+ concentrations are essential for many cellular processes, including the uptake of ammonia, CO_2, HCO_3^- and phosphate (ESPIE et al. 1988, FERNÁNDEZ and AVENDAÑO 1993). The cleavage of water in photosystem II has proved to be another Na-dependent process besides carbon transport in cyanobacterial photosynthesis (ZHAO and BRAND 1988). The essential role played by low Na^+ concentrations in maintaining a functioning nitrogen fixation system has been known for some time, Na^+ ions apparently being needed for the activation of nitrogenase (THOMAS and APTE 1984). Moreover, the transport of Na^+ is also involved in regulating internal pH values in cyanobacteria (PADAN and VITTERBO 1988). However, a concentration of 10 mM is sufficient to satisfy most of these needs (THOMAS and APTE 1984).

In the case of *Synechococcus* sp. PCC 7942, Na^+ appears to be taken up by electrogenically driven diffusion through the membrane (RITCHIE 1992a). There is some evidence that special pores in the membrane are involved in Na^+ transport in *Anabaena* strains (APTE and THOMAS 1986). It has been shown that the active motility of marine isolates of *Synechococcus* WH8113 is energized directly by an Na^+ motive force (WILLEY et al. 1987). Na^+-dependent energy- supplying processes also appear to be implicated in other cyanobacteria under alkaline growth conditions (BROWN et al. 1990).

Halophilic strains of cyanobacteria require at least 0.4 M NaCl in the medium for growth and can tolerate over 2 M NaCl without complete growth inhibition (MACKAY et al. 1984). The very low concentrations of inorganic ions in the cytoplasm of salt-adapted cyanobacteria compared with the concentrations in the surrounding medium suggest that such ions are actively extruded. After adaptation to sea water conditions (about 500 mM NaCl), slightly elevated Na^+ concentrations of between 20 and 100 mM per litre cell volume have been found in various cyanobacterial strains; the concentrations found in freshwater medium before adaptation were 10-30 mM (BATTERTON and VAN BAALEN 1971, BLUMWALD et al. 1983, WARR et al. 1984, REED et al. 1985a, Gabbay-Azaria et al. 1992, HAGEMANN et al. 1994). Extremely low Na^+ concentrations of less than 1mM have been reported for a strain of *Aphanothece* in a medium containing 2 M NaCl (MILLER et al. 1976), whereas REED et al. (1984b) measured values of around 250 mM Na^+ for another isolate of the same species and various other halophilic strains under similar salt loads.

In cyanobacteria, the intracellular K^+ concentration is distinctly higher than in the surrounding medium, and this is obviously a result of active accumulation. K^+ uptake is probably driven by the transmembrane potential and is at least indirectly dependent upon ATP (PASCHINGER 1977a, PADAN and VITTERBO 1988, RITCHIE 1992a). This ion is even probably actively transported in *Anabaena variabilis* (REED et al. 1981). Elevated K^+ concentrations have often been measured in salt-adapted cells of various cyanobacteria, and there is obviously no dramatic change in the Na^+/K^+ ratio in such cases (BLUMWALD et al. 1983, REED et al. 1985b, HAGEMANN et al. 1994). However, K^+ concentrations remain relatively constant in *Spirulina* strains (GABBAY-AZARIA et al. 1992) or even decline markedly in cells adapted to higher salinities (VONSHAK et al. 1988a). Extremely pronounced K^+ accumulation (up to 2M) has been reported for an *Aphanothece halophytica* isolate from a Californian salt lake (MILLER et al. 1976). K^+ ions at such high concentrations contribute substantially to the osmotic potential of the cell and, therefore, can be considered an osmoprotective substance in the same way as in heterotrophic bacteria (CSONKA and HANSON 1991).

Intracellular concentrations of Cl^- have been studied in comparatively few species. In cyanobacteria, these ions are accumulated against the membrane potential gradient by means of an Na^+ dependent process involving the consumption of ATP (DEWAR and BARBER 1974, RITCHIE 1992b). Like the concentrations of

Na^+ and K^+, the Cl^- concentration increases during salt adaptation, but always remains lower than in the surrounding medium (REED et al. 1985b, INCHAROENSAKDI and TAKABE 1988, GABBAY-AZARIA et al. 1992).

Like osmoprotective substances, these ions help to reduce the osmotic potential of the cytoplasm in salt-adapted cells. However, it is extremely difficult to draw up a balance showing the contributions of the various solutes to the total potential since measured values scarcely exist for the osmotically accessible cell volumes of cyanobacteria, the total cell volume generally being used for reference. Moreover, NMR studies dealing with ion accumulation in heterotrophic bacteria have shown that some of the ions in the cytoplasm are bound and therefore immobile. These, naturally, would not appear in an osmotic balance (KOGUT 1991). Using total cell volume as a basis, summation of the osmotically active substances (ions and osmoprotective substances) has shown that salt-adapted *Synechocystis* sp. PCC 6714 achieves only about 40 % of the osmotic potential of the surrounding medium (REED et al. 1985a). To enable growth, each cell must possess a lower osmotic potential than its surrounding medium in order to maintain turgor and take up water. Using the ESR method, MEHLHORN and SULLIVAN (1988) detected a distinct reduction in the osmotic potential of the cytoplasm and positive turgor in salt-adapted cells of *Synechococcus* sp. PCC 6311.

Although relatively low ion concentrations are characteristic of salt-adapted cells, very high transient Na^+, K^+ and Cl^- concentrations have been measured during investigations into the kinetics of salt adaptation. In *Synechococcus* sp. 6311 and *Synechocystis* sp. 6714, for instance, a strong influx of NaCl is observed within one or two minutes after a salt shock, the intracellular NaCl concentration increasing to between 50 and 90% of that in the surrounding medium (BLUMWALD et al. 1983, REED et al. 1985a, b, NITSCHMANN and PACKER 1992). A temporary breakdown in membrane permeability is thought to be responsible (REED et al. 1985a), but an Na^+/H^+ exchange process also appears to be implicated (NITSCHMANN and PACKER 1992). The Na^+ ions that enter the cell are actively extruded again in a rapid process lasting

10 to 20 minutes (REED et al. 1985a, b, NITSCHMANN and PACKER 1992), whereafter a new, lower steady state Na^+ concentration is achieved during a longer extrusion phase (BLUMWALD et al. 1983, NITSCHMANN and PACKER 1992). In *Synechocystis* sp. PC 6714, the Na^+ is replaced by less toxic K^+ while the sodium ions are being extruded. The accumulation of the osmoprotective substances glucosylglycerol and sucrose starts at this stage, and the concentrations of K^+ and Cl^- decrease rapidly (REED et al. 1985b). However, in salt-adapted cells of *Spirulina subsalsa*, a salt shock of 1 M NaCl induced a very slow increase in the NaCl content of the cells at first. It peaked after 24 h, and the new steady state with a relatively low ion concentration was not achieved until after several days (GABBAY-AZARIA et al. 1992). Salt-shocked cells of the same species did not extrude ions under dark conditions (GABBAY-AZARIA and TEL-OR 1993). RITCHIE (1992a) measured higher intracellular Na^+ concentrations in salt-stressed cells of *Synechococcus* sp. PCC 7942 during darkness than in the light. Cyanobacteria are probably unable to supply sufficient energy for salt extrusion in darkness, and the large amount of energy needed for this transport process would be another reason for their lack of salt adaptability in the dark.

At first, the rapid ion flux almost completely compensates for the osmotic potential difference between the intracellular and external media and thus counteracts further loss of water. The variation of ion export in time shows fairly accurately how long the accumulation of osmoprotective substances takes, the synthesis of the glycine-betaine system being particularly slow. This is possibly another reason for the relative slowness of ion regulation in *Spirulina subsalsa*.

2.2.6 Ion transport mechanisms

Most investigations into active Na^+ transport mechanisms have been conducted on various *Synechococcus* strains (*Anacystis nidulans*). This is probably a secondary active transport process involving a reversible Na^+/H^+ antiporter system which uses the H^+ gradient (proton motive force) across the cytoplasm membrane to export Na^+. Evidence of this transport mechanism has been furnished by measurements of the exchange of Na^+ and H^+ in the presence of various ATPase inhibitors and of respiratory electron transport (PASCHINGER 1977a, BLUMWALD et al. 1984, MOLITOR et al. 1986, SCHERER et al. 1988, PESCHEK et al. 1988, NITSCHMANN and PACKER 1992, PADAN and SCHULDINER 1993). So far, no definite proof has been found for the existence of a transport protein for the Na^+/H^+ antiporter in cyanobacteria, although one has been found in heterotrophic bacteria. Elevated levels of Na^+/H^+ antiporter activity have been reported in salt-adapted cells of *Synechococcus* (BLUMWALD et al. 1984, MOLITOR et al. 1986, NITSCHMANN and PACKER 1992).

The proton gradient needed across the cytoplasm membrane for Na^+ export can be generated in three ways: by H^+-transporting ATPases, by H^+ displacement during the respiratory e^- transport using cytochrome oxidase, and by reversible H^+-ATPases of the F_0/F_1 type (NITSCHMANN and PESCHEK 1985). So far, reversible H^+-ATPase has not definitely been detected in the cyanobacterial cytoplasm membrane, but unidirectional H^+-ATPases and cytochrome oxidases have and appear to be implicated in supplying energy for Na^+ extrusion. The activities of both enzymes are greatly elevated in various cyanobacterial strains after salt adaptation, i.e. under environmental conditions requiring more energy for ion export (MOLITOR et al. 1986, PESCHEK et al. 1988, NITSCHMANN and PACKER 1992, GABBAY-AZARIA et al. 1992, JEANJEAN et al. 1993, GABBAY-AZARIA et al. 1994). The contribution of respiratory e^- transport to the generation of the H^+ gradient for Na^+ transport increases from 15 to 50% in salt-adapted cells of *Synechococcus* sp. PCC6311 (NITSCHMANN and PACKER 1992). MOLITOR et al. (1990a, b) described how cytochrome oxidase activity on the cytoplasm membrane of *Anacystis nidulans* started increasing immediately after a salt shock of 0.4 M. It rose to three times its original level during the four-day adaptation phase, whereas its activity on the thylakoid membrane remained unchanged.

Increased cytochrome oxidase activity is reflected in increased respiration (O_2 consumption in the dark). Respiration is stimulated in *Anacystis nidulans* immediately after a salt shock (PASCHINGER 1977b, PESCHEK et al. 1988), and is also distinctly higher than normal in salt-adapted cells (PESCHEK et al. 1988, MOLITOR et al. 1990a, b).

Above-normal respiration values have also been described as characteristic of salt-stress in several other cyanobacterial strains (VONSHAK et al. 1988a, JEANJEAN et al. 1990, GABBAY-AZARIA et al. 1992, NITSCHMANN and PACKER 1992). Besides cytochrome oxidase, the activity of glucose-6-phosphate dehydrogenase, a key enzyme in sugar decomposition in the pentose phosphate cycle, is also elevated in salt-adapted cyanobacteria (GABBAY-AZARIA et al. 1989, HAGEMANN et al. 1989a, ERDMANN et al. 1992).

Conclusive proof that unidirectional H^+-ATPases are implicated in salt adaptation of *Synechococcus* sp. PCC7942 has been obtained by cloning the genes for two such enzymes. Both salt-sensitive and osmosensitive phenotypes were produced after inactivation of the genes coding for those ATPases (KANAMARU et al. 1993). The genes coding for cytochrome oxidase subunits in *Synechocystis* sp. PCC6803 have also been cloned recently, but so far none of their mutations has been studied for possible effects on salt tolerance (ALGE and PESCHEK 1993).

Elevated NaCl *in vitro* concentrations are known to stimulate both enzymes directly in *Spirulina subsalsa,* besides increasing the amount of cytochrome oxidase and an H^+-ATPase (GABBAY-AZARIA et al. 1992, 1994). Cytochrome oxidase activity in salt-adapted cells of *Synechocystis* sp. PCC 6803 is twice as high as in non-adapted cells, but the cell enzyme protein content is only slightly elevated (JEANJEAN et al. 1993). Key enzymes for the synthesis of some osmoprotective substances (mannitol, glucosylglycerol) can also be activated directly by NaCl (HAGEMANN and ERDMANN 1994), so enzymes that are essential for salt adaptation respond to a salt shock with increased activity, and those serving "normal" metabolism are simultaneously inhibited owing to their lower salt tolerance.

RITCHIE (1992a) does not support the view that Na^+ export is mediated by a proton driven antiporter (Na^+_{out}/H^+_{in}). His bioenergetic analysis of ion transport in *Synechococcus* sp. PCC 7942 indicates that Na^+ transport is driven by a primary Na^+-ATPase (three Na^+ per ATP cleavage), whereas the Na^+/H^+ antiporter serves the uptake of Na^+ and export of H^+ as part of the pH regulation mechanism (RITCHIE 1992a).

REED et al. (1985a) also reported increased Na^+ export activity in salt adapted cells of *Synechocystis* sp. PCC 6714. In this case, Na^+ extrusion was coupled to its ATP-dependent replacement by K^+, and REED et al. (1985a) therefore suggested that an Na^+/K^+-ATPase might be implicated. The exchange of K^+ for Na^+ and the implication of an ATP consuming process have also been described for *Agmenellum quadruplicatum* (BATTERTON and VAN BAALEN 1971). On the other hand, since it has not been shown that the ATPase activity associated with ion export can be inhibited by ouabain in cyanobacteria, it seems unlikely that a primary Na^+/K^+-ATPase is involved.

2.2.7 Effects on photosynthesis

In photo-autotrophic cells, photosynthesis usually correlates quite well with the growth rate. Since high salinities often inhibit cyanobacterial growth, the growth rate of cyanobacteria adapted to high salinities, unlike their respiration, is generally depressed (HAGEMANN et al. 1989b, TEL-OR 1980, BLUMWALD and TEL-OR 1984, VONSHAK et al. 1988a, ERDMANN et al. 1992, RAI and ABRAHAM 1993, SCHUBERT et al. 1993). The photosynthetic activity of *Synechocystis* sp. PCC 6803 and *Spirulina platensis* decreases quickly and drastically during the first few hours following a salt shock, but has almost reached its new steady state after 12 h (VONSHAK et al. 1988a, ERDMANN et al. 1992). In contrast, in *Synechococcus* sp. PCC 6311 stimulated photosynthetic activity was measured 24 h after exposure to salt shock although growth had been inhibited during the intervening period (BLUMWALD and TEL-OR 1984).

Apart from affecting overall photosynthetic activity, changes in salinity also lead to adjustments in the photosynthetic characteristics of cells of *Synechocystis* sp. PCC 6803. The PSII/PSI ratio is lower after salt adaptation (SCHUBERT and HAGEMANN 1990), mainly owing to an increase in PSI activity, but also to an increase in PSI content (JEANJEAN et al. 1993). As in the case of respiration, the increase in PSI activity is greater than the increase in the amount of PSI. The higher PSI activity is used mainly to increase the cyclic e^- transport (JEANJEAN et al. 1993), which might be a source of additional energy for ion extrusion by means of ATPases. Changes have also been found in pigmentation (RAI 1990, RAI and ABRAHAM 1993) and the transmission of energy between the photosystems and also in the phycobilisomes on PSII (SCHUBERT et al. 1993, JEANJEAN et al. 1993).

Carbon fixation decreases in *Aphanothece stagnina* (RAI 1990) and *Synechocystis* sp. PCC 6803 in proportion to the decrease in O_2 production (HAGEMANN et al. 1989b). RuBisCO activity is depressed in salt-loaded *Spirulina platensis* and *Aphanothece stagnina* cells, whereas PEP-carboxylase activity increases (TEREKHOVA et al. 1986, RAI 1990). The synthesis of osmoprotective substances leads to striking changes in the pattern of ^{14}C fixation (ERDMANN 1983, ERDMANN and SCHIEWER 1985). Osmoprotective substances predominate among the soluble fixation products, and increased carbon incorporation by β-carboxylation has been observed (ERDMANN and SCHIEWER 1985). In contrast, in *Aphanothece halophytica,* both CO_2 fixation and the RuBisCO content increase under high salt loads (TAKABE et al. 1988, RAI 1990).

2.2.8 Salt-induced changes in gene expression - stress protein synthesis

The bacterial metabolism is regulated principally at the transcription level, i.e. by the activation or repression of the expression of certain genes to suit the needs of cell growth. In view of the manifold physiological and biochemical changes that take place in salt-stressed cyanobacteria, it is scarcely surprising that first signs of a change in gene expression have also been reported.

Changes in the pattern of gene expression can be described fairly simply by characterizing the bacterial synthesis of protein under various growth conditions, this approach covering effects at both the transcription and translation levels. In cyanobacteria exposed to salt, osmotic or drought stress, three basic groups of protein can be identified by their response to environmental factors: 1) proteins whose expression remains relatively unaffected by variations in environmental factors ("household" proteins); 2) proteins whose synthesis is turned off or inhibited specifically when environmental conditions change; and 3) proteins whose synthesis is increased or induced specifically under unfavourable environmental conditions. The members of group 3) are called stress proteins and are of special interest because one may assume they perform certain functions during adaptation which permit the cells to achieve optimum growth in the changed environment.

Many cyanobacterial strains from various biotopes and with various salt tolerances synthesize the same stress proteins in response to salt stress (APTE and BHAGWAT 1989, HAGEMANN et al. 1990, 1991, BORBELY et al. 1990, HERSHKOVITZ et al. 1991b). Close study of the kinetics of stress protein synthesis have revealed that overall protein synthesis is strongly depressed immediately after the addition of salt to the medium (HAGEMANN et al. 1990) and that stress proteins predominate among the newly synthesized proteins during the first few hours of stress (APTE and BHAGWAT 1989, HAGEMANN et al. 1991). Normal protein synthesis is then gradually resumed as osmoprotective substances accumulate during the further adaptation process. Certain stress proteins are found only during the early stages of adaptation, i.e. their expression or induction is only transiently stimulated, whereas the synthesis rates of others are also, or only, high in salt-adapted cells (APTE and BHAGWAT 1989, HAGEMANN et al. 1991, 1994; Fig. 4). Salt-adapted cells of *Synechocystis* sp. PCC 6803 not only synthesize some proteins more intensively; they also accumulate them (HAGEMANN et al. 1994).

Comparison of the stress proteins synthesized by various cyanobacterial strains shows that their numbers and molecular weights vary from one strain to another (Table 1), proteins with a low molecular weight predominating. Most stress proteins are soluble, but several of them are also found on the cytoplasmic membrane (APTE and BHAGWAT 1989, HERSHKOVITZ et al. 1991, HAGEMANN et al. 1994).

The synthesis of several salt stress proteins produced by *Anabaena* strains is also stimulated by osmotic stress (BHAGWAT and APTE 1989, FERNANDES et al. 1993). However, since other proteins are synthesized specifically in response to salt or osmotic stress, the pathways for sensing and transmitting the stimulus obviously differ, although the responses are partly identical (FERNANDES et al. 1993).

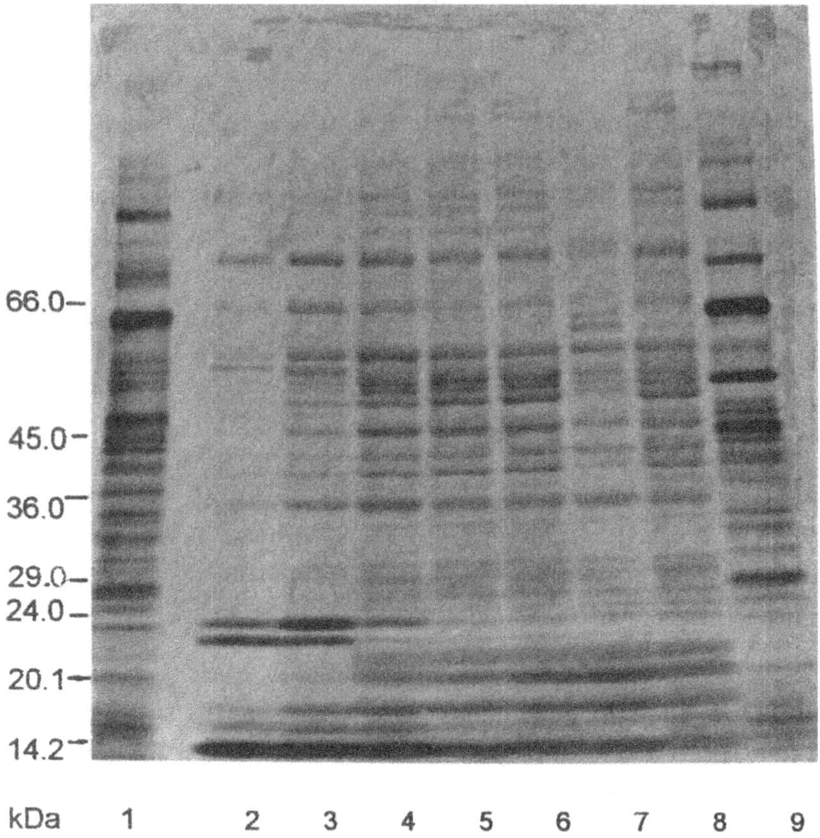

Fig. 4. Variation in time of salt stress protein synthesis in *Synechocystis* sp. PCC 6803 after a shock of 684 mM NaCl (1, 9 - controls; 2- 1h; 3- 2h; 4- 4h; 5- 6h; 6- 8h; 7- 10h after salt shock; 8- salt adapted) (HAGEMANN et al. 1991)

A similar response pattern has been found in drought-stressed cyanobacteria: both drought and salt stress stimulate the synthesis of various proteins, of which some are regulated exclusively by drought (HERSHKOVITZ et al. 1991b). The most studied drought-tolerant cyanobacterium is *Nostoc commune*, a species found in various polar, tropical and temperate biotopes. *N. commune* responds to repeated desiccation and moistening by synthesizing drought stress proteins which account for most of the soluble protein fraction and are accumulated extracellularly in the glycan envelope (SCHERER and POTTS 1989, HILL et al. 1994). Released when the cells are moistened, they constitute a considerable loss of nitrogen. The

Table 1. Comparison of molecular weights (kDa) of proteins synthesized in salt-loaded cells of various cyanobacterial strains (total protein extracts labeled with ^{35}S-methionine were separated by SDS polyacrylamide gel electrophoresis)

Anabaena sp. L31[1,2]	*Anabaena torulosa*1	*Synechocystis* sp. PCC 6803[3,4]
82.0[H]	79.0	72.0
66.0	54.0	63.5[H]
41.5	49.0	50.3
41.0	46.5	30.5[H]
38.5	32.0	28.5
35.0	29.0	24.7[H]
25.0	28.0	22.2
24.0	26.5	19.2
23.0[H]	21.5	16.9
21.0	19.5	14.2[H]
20.0	18.5	12.5
19.0[H]	17.0	
18.5		
18.0		

[1]APTE and BHAGWAT (1989); [2]BHAGWAT and APTE (1989); [3]HAGEMANN et al. (1991); [4]HAGEMANN et al. (1994); [H]Rate of synthesis also stimulated by heat shock

three drought stress proteins so far identified in *N. commune* (33, 37 and 39 kDa) possess very similar amino acid sequences and are therefore probably produced by proteolysis of a common protein precursor. They are aggregated with a UV A/B-absorbing pigment and account for up to 10% of the dry weight. Thus they confer protection against UV irradiation as well as desiccation (HILL et al. 1994).

Like interactions between salt, osmotic and drought stress, heat shock is also able to induce a few salt stress proteins (Table 1, BHAGWAT and APTE 1989, BORBELY et al. 1990, HAGEMANN et al. 1991). Stress proteins found after salt loading can therefore be grouped into general stress proteins (response to both salt and heat shock), general water stress proteins (response to salt, osmotic and drought stress, respectively) and those stress proteins that are specific to a given stressor.

Apart from their molecular weights, the sequence in which they are synthesized during adaptation and the effects of interactions between stressors on their expression, little is known about the functions of most stress proteins. Those induced even by heat shock probably have a non-specific protective (chaperon) function for macromolecules in the presence of high ion concentrations.

Stress proteins that are permanently and strongly expressed, probably play some direct role in the adaptation process. Various authors have reported an increase in the amount of cytochrome oxidase subunits (MOLITOR et al. 1986, PESCHEK et al. 1988, GABBAY-AZARIA et al. 1992). A 100 kDa protein that cross-reacts with antibodies against H^+-ATPase accumulates on the cytoplasm membrane of *Spirulina subsalsa* (GABBAY-AZARIA et al. 1994), and a 40 kDa protein with immunological similarities to the dehydrins of higher plants is expressed in various filamentous cyanobacterial strains in response to osmotic stress (CLOSE and LAMMERS 1993). A protective effect on the cell structure has been attributed to the water stress protein synthesized by *Nostoc commune* (SCHERER and POTTS 1989).

Little is known about the genetics of salt adaptation in cyanobacteria. APTE and HASELKORN (1990) have shown that the transcription of many genes is salt-regulated in *Anabaena torulosa*. Surprisingly, most of the salt-regulated genes in this species are lacking in the less salt-tolerant strain *Anabaena* sp. L31 or bear only a slight similarity to the genes of the latter. It has been estimated that at least 10 % of the *Anabaena* genome may be differentially expressed owing to salt regulation. Many of the salt-regulated genes have been cloned in *Escherichia coli*, and the increased expression of a salt-stress protein of *Anabaena* in *E. coli* has also been induced by increasing the salinity (APTE and HASELKORN 1990).

Several salt-defective mutants of the cyanobacterium *Synechocystis* sp. 6803 have been isolated and are being used to screen for salt tolerance genes (JEANJEAN et al. 1990, HAGEMANN and

ZUTHER 1992). The gene concerned has been isolated from one of these mutants and sequenced. Designated stpA (salt tolerance protein), the gene bears no clear resemblance to known sequences and, therefore, nothing definite can be said regarding its function. However, since the mutant exhibits pleiotropic defects, a complete lack of the ability to synthesize glucosylglycerol, depressed respiration and low cytochrome oxidase activity, the gene probably codes for a regulator protein (ONANA et al. 1994).

A freshwater cyanobacterium *Synechococcus* sp. PCC 7942 has been transformed with the *Escherichia coli bet* gene clusters (*bet*A, betB *bet*F and *bet*T). The genes were transferred to the chromosome of *Synechococcus* sp. PCC 7942, was demonstrated by northern blot analysis as well as by detection of glycine betaine in cells supplemented with choline. The transformant accumulated glycine betaine to a concentration of 45 mM that resulted in stabilisation of the photosynthetic activities, higher phycobilisome contents and general acquisition of resistance to salt stress (NOMURA et al. 1995).

2.2.9 Salt-induced changes in the membrane structure

The cyanobacterial cell has three membrane systems, the external, cytoplasm and thylakoid membranes. Few descriptions of the external membrane of cyanobacteria have been published. Like that of other gram-negative bacteria, it possesses pores with a certain degree of transport selectivity (HANSEL et al. 1994). The literature contains no work regarding the effect of varying salinities on the regulation of cyanobacterial pores. The permeability of the external membrane of *E. coli* is known to be strictly regulated by the expression of genes coding for pore-forming proteins (ompF and ompC) (CSONKA and HANSON 1991).

The cytoplasma membrane bounds the cell interior from the surrounding medium. Unlike the thylakoid membrane, the properties of which change only insignificantly during salt adaptation although the arrangement of the thylakoids in salt-stressed *Microcystis firma* cells differs markedly from the normal arrangement (SCHIEWER and JONAS 1977), the cytoplasm membrane plays a decisive role in salt adaptation. This is shown by comparing several properties of cytoplasm membranes from salt-adapted cells with those from control cells.

In salt adapted cells of *Synechococcus* sp. PCC 6311, the permeability of the membrane is depressed and its stability enhanced; plasmolysis within the cell can therefore take place only when the salt or sucrose concentration of the external medium is very high (BLUMWALD et al. 1983). In salt-adapted cells of *Anacystis nidulans, Synechococcus* sp. 6311 and *Aphanothece halophytica* the enhanced stability of the cytoplasm membrane is due to an increase in the proportion of longer chained saturated fatty acids and phospholipids (MOLITOR et al. 1990b, KHOMUTOV et al. 1990, RITTER and YOPP 1993). Heterotrophic eubacteria also characteristically stabilize their membranes under hypersaline conditions by changing its lipid and fatty acid composition (RUSSEL 1989, THIEMANN and IMHOFF 1991). The protein content of the cytoplasm membrane is also elevated (KHOMUTOV et al. 1990). Electron microscopic studies have shown that the density of protein particles in salt-adapted *Anacystis nidulans* cells increased, two structures being shown that correlate well with the increased cytochrome oxidase and Na^+/H^+-antiporter activity (MOLITOR et al. 1990a).

The salt tolerance of cyanobacteria is not affected exclusively by long term changes in their membrane properties. It is also strongly influenced by the permeability of the membrane to Na^+ and other ions. Enhanced salt tolerance owing to reduced permeability to ions has been observed in various strains of *Anabaena,* the addition of various nitrogenous compounds markedly reducing Na^+ uptake (APTE et al. 1987, REDDY et al. 1989, RAI and ABRAHAM 1995). Besides ammonium and nitrate, various amino acids have proved to be effective as nitrogen sources for reducing influxes of Na^+ (REDDY et al. 1989, THOMAS and SHANMUGASUNDARAM 1991). The Na^+ influx decreases very quickly after the addition of these nitrogenous compounds, possibly due to depolarization of the membrane since other ways of reducing membrane polarization also reduces its permeability to ions (APTE et al. 1987).

2.2.10 Effects of salt stress on nitrogen metabolism

Many cyanobacteria are able to fix atmospheric nitrogen. In many cyanobacteria, this has proved to be much more sensitive to salt loads than photosynthesis (TEL-OR 1980, DUBOIS and KAPUSTKA 1981, THOMAS and APTE 1984, MOORE et al. 1985, NIVEN et al. 1987, FERNANDES et al. 1993).

In two *Anabaena* isolates with different salt tolerances, the effects of NaCl stress were clearly distinguishable from pure osmotic stress caused by the addition of sucrose to the medium. In contrast to the strong inhibitory action of the salt stress, osmotic stress caused no inhibition over a wide range of concentrations and even stimulated N_2-fixation (FERNANDES et al. 1993). On the other hand, DUBOIS and KAPUSTKA (1981) demonstrated a distinct negative effect of increasing polyethylene glycol concentrations (a pure osmotic stress) on acetylene reduction, which they used as a measure of nitrogenase activity in the experiment. In drought-stressed *Nostoc commune* cells, nitrogenase activity is completely inhibited within 30 minutes (POTTS et al. 1984). The enzyme remains intact during short term desiccation so the nitrogen-fixing ability is quickly recovered when the cells are moistened (PEAT et al. 1988). However, after many years of desiccation, the cells needed much longer duration to regenerate their nitrogen-fixing ability than to recover their respiratory and photosynthetic capacity because no nitrogen-fixation was observed until they started forming new heterocysts (SCHERER et al. 1984).

Little is known about the causes of the low salt tolerance of nitrogen-fixation. However, ferredoxin-NADP-reductase (FNR) activity in filamentous cyanobacteria is also extremely sensitive to salt, and a lack of reducing agents and ATP is therefore under discussion as a possible cause (TEL-OR 1980). *In vitro* analysis of FNR activity in *Anabaena variabilis* in the presence of increasing NaCl concentrations also revealed a marked salt sensitivity, but it was much less pronounced than the *in vivo* salt sensitivity of nitrogenase activity (NIVEN et al. 1987). Sudden increases in salinity change the permeability of the membranes, and increased O_2 concentrations as well as inflowing ions might therefore lead to a denaturation of nitrogenase in the heterocysts (NIVEN et al. 1987). FERNANDES et al. (1993) showed that the degree of nitrogen-fixation inhibition due to NaCl stress can be reduced by the simultaneous addition of ammonium or nitrate to the medium. RAI and ABRAHAM (1995) noted that the uptake of Na^+ was minimum when nitrate was simultaneously available to the cells of *Anabaena doliolum*, indicating the interaction of nitrate with the Na^+ carrier. Na^+ efflux was maximum in N_2, ammonia, urea and nitrate in decreasing order. This led to the conclusion that Na^+ influx plays the critical role in salt tolerance rather than its efflux (RAI and ABRAHAM 1995). According to REDDY et al. (1989), in the presence of these nitrogenous compounds less energy (ATP) is needed for active ion extrusion and more remains available for nitrogen-fixation (FERNANDES et al. 1993). However, the long term presence of nitrogen sources depresses nitrogen-fixation.

Apart from nitrogen-fixation, few processes directly linked to nitrogen metabolism have been studied for salt-induced effects. Glutamine synthetase has been used as a model enzyme in several *in vitro* studies dealing with the effects of increasing salt concentrations, but no negative effects have been detected at physiologically relevant concentrations (TEL-OR 1980, WARR et al. 1984, 1988).

There is some evidence for a reduction of nitrate uptake by salt-loaded cyanobacteria, and it has been suggested that this might be responsible for the increased accumulation of glycogen in salt-adapted cyanobacteria (POTTS et al. 1983, LEFORT-TRAN et al. 1988, KERBY et al. 1990, ERDMANN et al. 1992). Further characteristics of nitrogen-deficient cultures such as reduced phycocyanin and protein levels have also been found in salt-loaded cyanobacteria (BLUMWALD and TEL-OR 1982, SCHUBERT et al. 1993, HAGEMANN et al. 1994).

Amino acids are scarcely implicated in the salt adaptation of cyanobacteria. It has been reported that the accumulation of proline by overproductive mutants leads to NaCl tolerance, a close correlation existing between the amount of surplus proline and the degree of salt tolerance (RICCARDI et al. 1983). Like inorganic nitrogen sources, the presence of some amino acids (alanine, aspartate, glutamate, glutamine, but not proline) in the extracellular medium reduces the flux of Na^+ through the cytoplasm membrane and thus enhances salt tolerance (REDDY et al. 1989, THOMAS and SHANMUGASUNDARAM 1991).

A major readjustment of nitrogen metabolism can be expected in cyanobacteria that accumulate glycine betaine. The accumulation of this substance is believed to bind large quantities of nitrogen otherwise used for other processes. RAVEN (1985) compared the accumulation of nitrogen-free osmoprotective substances (mannitol and sorbitol) with the accumulation of some containing nitrogen (proline and glycine betaine) in a kind of cost-benefit analysis. The energy (photons) and carbon inputs were only about 10% higher for the substances containing nitrogen than for those containing none, the main difference lying in the very large amount of nitrogen needed to synthesize the former. The inputs needed to provide nitrogen for use in osmoprotective substances more than doubles under salt stress conditions (RAVEN 1985).

2.3 Temperature

Within a certain temperature range, the effects of temperature on all metabolic processes produce an optimum curve bounded by the temperature tolerances of the various enzymes. Like growth, all metabolic processes have a species-specific minimum, optimum and maximum temperature. In the field, cyanobacteria can survive over a wide range of temperatures. In cultures, most strains exhibit optimal growth between 30 and 35°C, but thermophilic strains isolated from hot springs have a much higher growth optimum and tolerate up to 65°C.

2.3.1 Heat stress

A response to heat shock can be elicited in cyanobacteria by suddenly increasing the growth temperature by 10°C. The response is very similar in all organisms, from archaebacteria, eubacteria and plants to animals, and includes strongly inhibited growth and inhibition of most protein synthesis. However, its chief characteristic is the massive expression of several heat stress specific proteins.

2.3.1.1 *Heat shock proteins (hsp)*

The synthesis of heat stress proteins has been studied in several cyanobacteria during the past 10 years. They have been detected in varying numbers in both unicellular and filamentous cyanobacterial strains by means of labelling with radioactively labeled amino acids and subsequent one or two-dimensional electrophoresis (BORBELY et al. 1985, 1990, SURANYI et al. 1987, BHAGWAT and APTE 1989, WEBB et al. 1990, HAGEMANN et al. 1991, NICHOLSON et al. 1991, BLONDIN et al. 1993, Fig. 5).

The synthesis of hsp starts immediately after heat exposure and can be inhibited with rifampicin (BORBELY et al. 1990). However, not

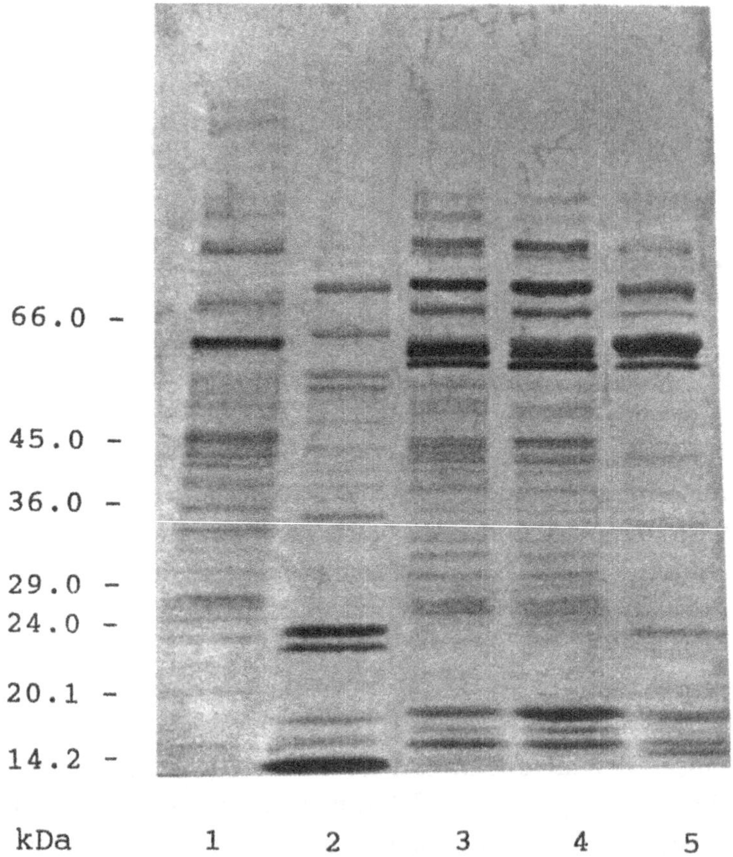

Fig. 5. Comparison of stress protein synthesis induced by salt and heat shock in *Synechocystis* sp. PCC 6803 (1 - control; 2 - 2 h salt shock of 684 mM NaCl; 3 - 40°C; 4 - 43°C; 5 - 46°C)

all hsp are exclusively heat-induced, many of them being constitutively expressed and merely being synthesized much more rapidly after heat shock. Their constitutive expression indicates that they also perform essential functions in the normal cellular metabolism (PARSELL and LINDQUIST 1993). Following the initial heat exposure, the increased rate of hsp synthesis characteristically tails off after a few hours (LEHEL et al. 1992, YURA et al. 1993). Comparison of the molecular weights and numbers of hsp from various cyanobacterial strains shows that the strains differ in both respects, having only a 65 kDa protein in common (Table 2).

Table 2. Comparison of heat shock proteins of different cyanobacteria with those of *E. coli*

	E. coli[a]	*Synechococcus* sp.		*Anabaena* sp.		*Synechocystis* sp.
		6301[b]	7942[c]	7120[d]	L31[e]	6803[f]
lon	94.0	91.0	80.0	110.0	92.0	88.0
htpM	84.1	78.0	70.0	87.0	82.0	75.0
htpG	71.0	65.0	65.0	74.0	75.0	65.0
rpoD	70.3	64.0		67.0	65.0	58.0
dnaK	69.1	58.0	58.0	65.0	32.0	33.0
groEL	62.9	49.0		35.0	23.0	23.0
lysU	60.5		39.0	24.0	19.5	17.5
htpI	48.5		28.5	21.5		14.2
dnaJ	41.0	24.0	23.5	18.0		13.8
htpH	33.4	18.0	18.0	17.0		
grpE	25.3	14.0		15.6		
htpL	21.5			14.3		
htpO	21.0			13.0		
htpE	14.7					
htpN	13.5					
groES	10.7					
htpK	10.1					

[a]NEIDHARDT and VAN BOGELEN 1987, bBORBELY et al. 1985, [c]WEBB et al. 1990, [d]BORBELY et al. 1990, [e]BHAGWAT and APTE 1989, [f]HAGEMANN et al. 1991

Similarities between the molecular weights of cyanobacterial hsp and those of *E. coli* suggest that functional similarities also exist. Several hsp act as molecular chaperons in cells of widely varying origin. Chaperons are able to protect other proteins against denaturation, to assist in transport and folding and to repair damaged proteins by refolding them. In addition, the hsp include proteases that specifically decompose irreversibly damaged proteins in heat-treated cells (PARSELL and LINDQUIST 1993).

The gene coding for the chaperon groEL in *Synechococcus* sp. PCC7942 and PCC6301 and in *Synechocystis* sp. PCC 6803 has been isolated and sequenced (COOKSON et al. 1989, WEBB et al. 1990, CHITNIS and NELSON 1991). Not far from the groEL sequence, a groES sequence closely resembling that gene in *E. coli* has been identified in *Synechococcus*. Both genes in *Synechococcus* form an operon and are transcribed together as 2.4 kb mRNA (WEBB et al. 1990). The *Synechocystis* groEL gene, in contrast, is transcribed alone as monocistronic RNA (CHITNIS and NELSON 1991). The gene for another chaperon, dnaK, has also been isolated from *Synechocystis* sp. PCC 6803. Like groEL, the gene coding for dnaK is very similar to sequences in *E. coli* and also in plants. It is present on the chromosome as a single copy and is also expressed as monocistronic RNA (CHITNIS and NELSON 1991).

The increase in the expression of groEL/ES and dnaK is due to increased transcription. Northern blot experiments revealed a large increase in the amount of RNA for these genes a few minutes after heat shock. In *Synechococcus* sp. PCC7942, a maximum representing about 120-fold increase in RNA was measured 20 minutes after heat shock (30-45°C), whereas the amount of protein increased only by a factor of 10. Expression was only slightly elevated 60 minutes after the heat shock (WEBB et al. 1990). The increased transcription of the groEL and dnaK genes in *Synechocystis* sp. PCC6803 was also only transient. A 70-fold increase in the amount of RNA was observed 60-120 minutes after heat treatment (20-42°C), but the level had almost returned to normal again after three hours (CHITNIS and NELSON 1991).

The promoter of the groEL operon of *Synechococcus* sp. PCC7942 has been mapped exactly. Its sequence exhibits great similarity to the consensus sequence of promoters for *hsp* genes in *E. coli,* but differs considerably from promoter sequences for *Synechococcus* genes that are not heat regulated (WEBB et al. 1990). In *Synechocystis* sp. PCC 6803, the sequences coding for groEL and dnaK are also preceded by regions that strongly resemble typical *hsp* gene promoters in *E. coli* (CHITNIS and NELSON 1991). In *E. coli,* the promoter of *hsp* genes is identified by an alternative sigma factor which is activated in cells exposed to heat shock and specifically induces the more intensive expression of *hsp* (YURA et al. 1993). In view of the structural similarities between the promoters of genes coding for hsp in cyanobacteria and *E. coli,* it seems likely that an alternative sigma factor is also implicated in the cyanobacterial response to heat shock (WEBB et al. 1990), although there is no definite proof. The gene coding for the principal sigma factor has already been isolated from *Anabaena* sp. PCC 7120. The structure and function of this factor are very similar to those of the vegetative sigma factor of *E. coli* and *Bacillus subtilis* (BRAHAMSHA and HASELKORN 1991). Further sequences coding for putative sigma factors have been identified on the chromosome of *Anabaena* in other studies (BRAHAMSHA and HASELKORN 1992). Mutations of individual genes coding for the various sigma factors had no effect on environmentally controlled gene expression processes in pilot experiments (BRAHAMSHA 1993).

In *Synechocystis* sp. PCC 6803, the groEL protein is an oligomer of varying molecular weight and can dissociate into monomers depending on the ionic strength of the buffer. It may have various forms of organization bonded to various macromolécules (e.g., phycobilisomes) in the cell (LEHEL et al. 1992). In iron-deficient *Synechococcus,* the groEL protein has been found in association with the chlorophyll binding protein CPVI-4 (WEBB et al. 1990). In *Anabaena* sp. PCC 7120, the groEL protein has been detected in both vegetative cells and heterocysts, most of it being concentrated on the thylakoid membrane and in carboxysomes in vegetative cells (JÄGER and BERGMAN 1990). It is possible that this chaperon is implicated in the synthesis of RuBisCO in the carboxysomes because GOLOUBINOFF et al. (1989a,b) showed that it is involved in the assembly of RuBisCO in *E. coli* cells, in which this enzyme has also been expressed by cyanobacterial genes.

Indirect evidence that hsp are involved in the heat tolerance of cyanobacteria is furnished by the fact that short term heat shock treatment at sublethal temperatures and long term adaptation to higher temperatures increases the survival rate at normally lethal temperatures (BLONDIN et al. 1993, LEHEL et al. 1993). In experiments with *Synechocystis* sp. PCC 6803 culture, growth is accompanied by increased synthesis of the groEL protein even at 39°C. After heat shock, the increase in hsp expression in these cultures was greater than in cells adapted to 22°C (LEHEL et al. 1993). Since photosynthetic O_2 production and PSII activity were also less affected by the heat, the chaperons conceivably associate with thylakoid structures and protect their structures and function (LEHEL et al. 1993). This protective function has been demonstrated experimentally in *E. coli* mutants lacking the chaperon genes and therefore capable of growth only in a very narrow temperature range (PARSELL and LINDQUIST 1993). Overexpression of an hsp60 chaperon also leads to the protection and reactivation of heat labile enzymes in mitochondria of heat shocked yeasts (MARTIN et al. 1992).

2.3.1.2 *Membrane properties*

Long term adaptation to different temperatures induces changes in membrane composition. Adaptation to relatively high temperatures increases the proportion of saturated fatty acids in cyanobacterial membranes, thus enhancing their stability (SATO and MURATA 1980, MURATA 1989). The lipid composition of the thylakoid membrane probably also affects photosynthesis since certain lipid fractions are associated with photosynthesis proteins.

The effect of changes in unsaturated fatty acid content on the heat stability of photosynthesis has been studied directly on *Synechocystis* sp. PCC 6803 mutants with a defective fatty acid saturase (*des*A) gene. The stability of PSI, PSII and total photosynthetic activity after heat shock was found to be independent of the saturated and unsaturated fatty acid contents (GOMBOS et al. 1991). However, more recent studies have shown that the removal of two-fold unsaturated lipid molecules (corresponding to an increase in saturation) tends to reduce heat tolerance rather than increase it (GOMBOS et al. 1994).

MILLER et al. (1989) reported that adaptation of a thermophilic *Synechococcus* strain to high temperatures (38-58°C) increased the PSI/PSII ratio, but PSII activity remained relatively constant. This meant that more ATP could be provided by cyclic electron transport to meet the possibly greater energy demand for growth at high temperatures (MILLER et al. 1989). Compared to that of other nitrogen-fixing bacteria, nitrogen-fixation by cyanobacteria has a relatively high heat tolerance (APTE 1992).

2.3.2 Cold stress

Compared with other strains, laboratory cultures of *Anacystis nidulans* are remarkably sensitive to low temperatures. A few minutes of incubation at 5°C is lethal to about 50% of the cells (MURATA 1989). This is due to the lower proportion of polyunsaturated fatty acids in *Anacystis* than in other cyanobacteria (WADA et al. 1990).

So far, the changes taking place in the protein synthesis pattern of cold-shocked cyanobacteria have not been studied. However, in *E. coli* a group of cold induced proteins has been described that are analogous to heat shock proteins (JONES et al. 1987, GOLDSTEIN et al. 1990). Cyanobacteria can be expected to respond in a similar way to judge from changes in enzyme activity and gene expression. RADY et al. (1994) showed that in *Synechocystis* sp. PCC 6803, for instance, antioxidative enzyme (superoxide dismutase, catalase and gluthatione peroxidase) activity increased and the protein content decreased following cold treatment (36-22°C), whereas heat stress (36-43°C) had the opposite effect.

SATO (1992) reported that the amount of mRNA from about 10 genes found in *Anabaena variabilis* increased 2 h after the temperature had been reduced (38-22°C), so cold-induced proteins evidently also exist in cyanobacteria. Three of these cold regulated genes have been characterized. The sequence of the *lti*2 gene of *Anabaena* strongly resembles Ó-glucosyl transferase or α-glucosyl hydrolase. The role these proteins play in cold adaptation is still unclear. The expression of *lti*2 had increased by a factor of 40 one hour after exposure to cold and then gradually decreased again so there was no sign of enhanced expression in cells adapted to 22°C (SATO 1992). Two other genes organized in one operon were identified among the cold stimulated genes. They coded for a RNA binding protein (rbpA) and a ribosomal protein (rpsU). The amount of their mRNA had increased 5-10 fold by 2.5 h after the temperature had been lowered from 38 to 22°C (SATO 1994).

2.3.2.1Membrane properties

Adaptation to low temperature is accompanied by an increase in the proportion and degree of unsaturated fatty acids of all lipid classes in the cyanobacterial membrane (MURATA 1989). This reduces the temperature of the lipid phase transition (transition from the liquid-crystalline to the phase-separated state), thereby increasing the fluidity of the membrane. The phase transition temperature is higher for the thylakoid membrane than for the cytoplasmic membrane owing to the higher protein content of the former, but adaptation to low temperature reduces the transition temperature of both (MURATA 1989).

Work with *Anacystis* cells, has shown that the fatty acid composition of the cytoplasmic membrane is critical for cold tolerance (VIGH et al. 1985, GOMBOS and VIGH 1986). MURATA (1989) used this knowledge to elaborate a model of cold sensitivity in cyanobacteria: the transition of the cytoplasmic membrane into the phase-separated state is accompanied by a loss of permeability, so that metabolites flow out of the cell and photosynthesis is irreversibly disrupted.

Direct proof for this hypothesis was furnished recently by cloning a cyanobacterial fatty acid desaturase. Mutants of *Synechocystis* sp. PCC 6803 with defective unsaturated fatty acid synthesis in the ^{6}position (Fad6) or ^{12}position (Fad12) of C18 fatty acids were isolated. The Fad12 mutants exhibited depressed growth compared with the wild type at 22°C, but not at 34°C (WADA and MURATA 1989). The Fad12 mutants were used to isolate a gene coding for a fatty acid desaturase (desA) (WADA et al. 1990). The *Synechocystis* sp. PCC 6803 *des*A gene was transferred to *Anacystis nidulans* RS, where it was obviously expressed because an increase in poly-unsaturated fatty acids was found in the transformed cell. This lowered the temperature at which phase separation with the attendant loss of permeability took place, thereby enhancing cold tolerance (WADA et al. 1990).

The single mutant Fad6 exhibited no difference in low temperature growth. Deletion of the *des*A genes in Fad6 created a double mutant with an even lower polyunsaturated fatty acid content than either of the single mutants Fad6 and Fad12. The growth of this mutant was even slower at 22°C, and its sensitivity to photo-inhibition after cold shock (10°C) was much greater (WADA et al. 1992).

2.4 Light intensity

As photoautotrophic organisms, cyanobacteria need light as an energy source for carbon fixation and other metabolic activities. However, variations in light quality also acts as signals triggering cellular responses and differentiation (TANDEAU DE MARSAC and HOUMARD 1993). In their natural environment, cyanobacteria are exposed daily to light fluctuations ranging from darkness to supra-optimal intensities at around noon. Excessive light intensity leads to initially reversible, but ultimately irreversible inhibition and disruption of photosynthetic activity. However, effective adaptation strategies have evolved for optimizing the photosynthetic apparatus to suit the prevailing light conditions.

In addition, light has a regulatory effect on carbon metabolism. In the case of cyanobacteria, the enzymes serving the oxidative pentose phosphate cycle (ppc) (decomposition of sugars) are always inhibited by light, whereas parts of the reductive ppc are stimulated (sugar assimilation). The activities of these enzymes are regulated by the reduction/oxidation status of the cytoplasm through the mediation of the thioredoxin and ferredoxin/thioredoxin system of the cell (ROWELL et al. 1988, CRAWFORD et al. 1988). However, the NADPH/NADP ratio may also be directly implicated (CSEKE et al. 1981). Other metabolic processes are sure to be affected besides those involved in sugar metabolism.

2.4.1 Pigmentation

An inverse relationship between pigment quantity and light intensity has often been observed in cyanobacteria. Cultivation at high light intensities leads to a reduction in chlorophyll and phycobiliprotein content, whereas the quantity of carotenoids remains relatively stable or even increases (ALLEN 1968, VIERLING and ALBERTE 1980, RAPS et al. 1983).

Chlorophylls are associated with photosystems, and the amounts in those systems are regulated to suit the amount of light. The amount of PSII remains relatively constant despite light intensity fluctuations. This is shown by the decrease in the PSII/PSI ratio from about 1 in week light to 0.25 in intense light (KHANNA et al. 1983, FUJITA et al. 1988, REUTER and MÜLLER 1993). For their model describing the regulation of the ratio between the photosystems, MURAKAMI and FUJITA (1991) assumed that the redox steady state of the b_6-f complex acts as a signal.

The phycobilipigments are contained in the phycobilisomes, the extrinsic light-collection complexes of PSII. Cyanobacteria possess various types of phycobilisome, and these are adapted to light variations in different ways (REUTER and MÜLLER 1993). Adaptation to strong light, for instance, can involve a reduction in the number of phycobilisomes and their associated PSII complexes (RAPS et al. 1985, FUJITA et al. 1990). The phycocyanin content of *Synechococcus* sp. PCC 7942 increases under weak light, and also under strong light in the presence of DCMU, but the synthesis of the D1 protein of PSII decreases. The rate of D1 synthesis might therefore be a regulator for light adaptation (KÖNIG 1990). Changes in the structure of the phycobilisomes have also been reported, adaptation to strong light sometimes being accompanied by the loss of whole rods (GROSSMAN et al. 1993). The phycocyanin/allophycocyanin ratio in *Synechococcus* sp. PCC 6301 decreases by around 50% under strong light (YAMANAKA and GLAZER 1981). The phycobilisomes of such cells therefore shrink owing to the shortening of their peripheral structures (BELKNAP and HASELKORN 1987, DE LORIMIER et al. 1992). Cyanobacteria that can synthesize phycoerythrin besides phycocyanin accumulate more phycoerythrin under weak light conditions, so that the composition and absorption properties of their phycobilisomes change (BRYANT 1982, BELKNAP and HASELKORN 1987, SWANSON et al. 1992, REUTER and NICKEL-REUTER 1993).

Since practically all genes coding for phycobilisome constituents have been identified, the molecular mechanisms of adaptation processes can now be studied (GROSSMAN et al. 1993, TANDEAU DE MARSAC and HOUMARD 1993). SWANSON et al. (1992) and DE LORIMIER et al. (1992) reported that under weak light conditions the amount of RNA for phycobiliproteins increases at the same time as the amount of pigment in *Anabaena* sp. PCC 7120 and *Synechococcus* sp. PCC 6803. On the other hand, elevated transcript levels for phycobiliproteins have been found in cyanobacteria exhibiting reduced phycobiliprotein levels after cultivation at high light intensities (TANDEAU DE MARSAC et al. 1988, BELKNAP and HASELKORN 1987). These seemingly contradictory findings suggest that translational control and the stability of phycobiliproteins play an important role in pigment regulation besides transcription.

Carotenoids are found in the cell wall and in both the cytoplasm and thylakoid membranes of cyanobacteria (REUTER and MÜLLER 1993). As antenna pigments, these are implicated in photosynthesis, but they are also protective pigments since they detoxify the singulet oxygen produced as a toxic by-product of photosynthetic water cleavage in PSII under intensive light. Several carotenoid binding proteins have already been identified in cyanobacteria, and the higher rate at which some are synthesized under high O_2 loads and intense light underscores their protective role (BULLERJAHN and SHERMAN 1986, BULLERJAHN et al. 1987, MASAMOTO et al. 1987).

Like carotenoids, various enzymes such as superoxide dismutase, catalase and peroxidase also eliminate intracellular radicals and therefore contribute to overcoming photoinhibition (STREB et al. 1993). GRILLI CAIOLA et al. (1991) have reported that superoxide dismutase activity is elevated in *Anabaena cylindrica* cells exposed to high light intensities. In *E. coli*, the expression of manganese-dependent superoxide dismutase from *Anacystis nidulans* enhances resistance to oxidative stress (LAUDENBACH et al. 1989, GRUBER et al. 1990).

2.4.2 Instability of photosystem (PS) II

Although PSI is very stable in cyanobacteria even when exposed to very high light intensities, PSII activity is quickly inhibited (YOKOYAMA et al. 1991). Turnover of the D1 protein in PSII is always rapid, and under light conditions causing photoinhibition levels of this protein decline because it is decomposed faster than the cell can replace it or reintegrate it into PSII (VONSHAK et al. 1988b, KRUPA et al. 1990, KANERVO et al. 1993).

Unlike the chloroplasts, in cyanobacteria the D1 protein is coded for by a gene family (*psb*A1 - 3) and therefore exists in two very similar forms (form 1 = *psb*A1, form 2 = *psb*A2 and *psb*A3) (TANDEAU DE MARSAC and HOUMARD 1993). The transcription of these genes and the amounts of the corresponding proteins synthesized under varying light conditions have been studied in *Synechococcus* sp. PCC 7942. The expression of *psb*A2 and *psb*A3 (D1 form 2) is enhanced by strong light, whereas under weak light conditions expression is restricted almost entirely to *psb*A1 coding for form 1, which accounts for most of the D1 protein (SCHAEFER and GOLDEN 1989a, b). Mutants of *Anacystis nidulans* R2 from which the *psb*A1 gene has been deleted contain only form 2 of the D1 protein; they are somewhat less liable to photoinhibition and exhibit a distinctly higher repair rate (KRUPA et al. 1990). SAMUELSSON et al. (1987) reported that, after adaptation to strong light (relatively high proportion of D1 protein form 2), *Anacystis* reaches photosynthetic saturation and probably also photoinhibition only at higher light intensities.

Cyclic electron transport around PSI and respiration may also help to protect PSII during photoinhibition since inhibition of these processes aggravates the damage caused by excessive light intensity (FORK and HERBERT 1993, SHYAM et al. 1993). Photosynthesis and respiration use the same redox carriers on the thylakoid membrane, and this ensures that the membrane is energized and the Cytb_6-f complex reduced when PSII is inhibited (MATTHIJS and LUBBERDING 1988).

2.5 Darkness

Cyanobacterial growth stops rapidly when the cells are transferred from a light to a dark environment. Macromolecule synthesis ceases (MARINO and ASATO 1986), and stored glycogen is consumed as an energy source. The expression of many genes, especially those implicated in photosynthesis, is light-dependent (TANDEAU DE MARSAC and HOUMARD 1993). The regulation mechanisms are gene and strain specific. The photosynthetic apparatus of the facultatively heterotrophic cyanobacterium *Anabaena variabilis*, for instance, still functions after cultivation in darkness for six months (MANNAN and PAKRASI 1993).

The transition to conditions of darkness is also accompanied by a change in protein synthesis patterns, and the synthesis of various proteins found in *Synechococcus* sp. PCC 6301 and *Chlorogloea fritschii* is induced or enhanced by darkness (SURANYI et al. 1987, FAULKNER 1988, BORBELY et al. 1990).

3 Signal recognition and transmission

It is not yet completely clear how a cell recognizes a stress situation and activates the appropriate adaptation mechanisms. Of the various known and possible signal pathways, we shall concentrate here mainly on signal transmission via messenger cascades and the two component system.

3.1 Messenger cascades

The key elements of this signal transmission pathway are low molecular signal molecules that form a chain and ultimately activate an enzyme by protein phosphorylation. The best known signal molecules are the second messengers cAMP, Ca^{2+} and highly phosphorylated nucleotides. cAMP is quite common in prokaryotes (BOTSFORD and HARMAN 1992) and is also found in cyanobacteria. Adenylate cyclase, the enzyme synthesizing cAMP, is present in *Anabaena cylindrica* and *Spirulina platensis*, but has not been found in *Synechocystis* sp. PCC 6803 (SAKAMOTO et al. 1991). The transfer of *Anabaena* cultures to a dark environment promotes cAMP synthesis, whereas light exposure quickly reduces cAMP levels (OHMORI et al. 1988). Like darkness, conditions leading to a reduced electrochemical membrane potential - a characteristic feature of energy deficiency states - appear to stimulate cAMP synthesis (OHMORI 1989). Extracellular cAMP added to dense *Spirulina platensis* cultures induces cell aggregation and even the formation of real mats (OHMORI et al. 1992).

Besides the second messenger cAMP, highly phosphorylated nucleotides acting in some unknown way as alarmones are also involved in the transduction of environmental signals. Several stress treatments induce cyanobacteria transiently to synthesize these highly phosphorylated nucleotides, which in heterotrophic eubacteria are implicated in the stringent response. Growth inhibition induced in *Anacystis nidulans* by reduced light intensity leads briefly to the rapid synthesis of guanidine tetraphosphate and pentaphosphate (ppGpp, pppGpp, MANN et al. 1975). Darkening of *Synechococcus* sp. PCC 6301 also induces the synthesis of these nucleotides (SURANYI et al. 1987, BORBELY et al. 1990). In both cases, RNA synthesis slows down at the same time as growth. However, prolonged heat treatment also leads to ppGpp accumulation in *Synechococcus* sp. PCC 6301 under both light and dark conditions (SURANYI et al. 1987, PALFI et al. 1991). A heavy metal treatment, in contrast, causes the accumulation of various bisnucleotide oligophosphates, among which adenine and guanidine oligophosphates predominate in various combinations. To some extent, the kinds of oligonucleotide phosphate accumulated even seem to be specific to certain heavy metals (PALFI et al. 1991).

3.2 Two component systems, protein phosphorylation

The two component system is a relatively simple signal transduction system and is almost paradigmatic for the environmental monitoring and adaptation activities of prokaryotic organisms.

As a rule, a two component system consists of a membrane bound protein kinase acting as a sensor and a regulator protein which is activated by phosphorylation and, in the simplest case, acts as a transcription factor (STOCK et al. 1989, SAIER 1993).

The genes for one two component system of *Synechococcus* sp. PCC 7942 have been cloned. Genes *sph*S and *sph*R code for a sensor kinase and a response regulator that are probably implicated in adaptation to phosphate deficiency since mutation of these genes prevents typical responses to phosphate deficiency, for instance the induction of alkaline phosphatase synthesis (AIBA et al. 1993). Phosphoproteins also act as transcription regulators during the response to variations in nitrogen availability. The gene *ntc*A, coding for a global nitrogen regulator and resembling other bacterial DNA-binding proteins has also been isolated from *Synechococcus* sp. PCC 7942 (VEGA-PALAS et al. 1992). However, DNA sequences that strongly resemble eukaryotic protein kinases and phosphatases have also been found in cyanobacteria besides genes and proteins similar to parts of typical bacterial two component systems (ZHANG 1993, POTTS et al. 1993).

Several environmentally induced protein phosphorylations in cyanobacteria have been described without reference to the activation pathways leading to them. In *Anabaena* sp. PCC 7120, for instance, phosphorylation appears to be involved in the regulation of carbon metabolism (MANN et al. 1991). A change between photoheterotrophic and autotrophic growth at various CO_2 concentrations (BLOYE et al. 1992) and salinities (HAGEMANN et al. 1993) induced changes in protein phosphorylation in *Synechocystis* sp. PCC 6803. The binding specificity of proteins for light regulated genes during the adaptation of *Calothrix* sp. PCC 7601 to various light qualities in the course of complementary chromatic adaptation is influenced by their phosphorylation status (SOBCZYK et al. 1993).

Relatively sudden changes in light quality lead to a corresponding redistribution of energy absorption between the two photosystems as part of the so called state transition (BIGGINS and BRUCE 1989). The cultivation of *Synechococcus* sp. PC 6301 under light that is absorbed predominantly by PSI or PSII respectively leads to changes in the phosphorylation status of a few proteins (SANDERS et al. 1989). A 13 kDa protein that is phosphorylated under PSII light but remains unchanged under PSI light has been isolated and partly sequenced. Its N-terminal sequence bears similarities to the PII protein (gene *gln*B) of *E. coli*, which is implicated in the regulation of glutamine synthetase activity to suit nitrogen availability (HARRISON et al. 1990). The *glnB* gene of *Synechococcus* sp. PCC7942 has now been cloned, and the covalent modification of its gene product confirmed. It has been shown that the post-translational modification of the PII protein can be suppressed regardless of light quality by adding ammonium to the medium. Therefore, the nitrogen metabolic pathway in cyanobacteria appears to be regulated by photosynthetic electron transport (TSINOREMAS et al. 1991). Many environmental factors influence photosynthesis and lead to changes in the redox status of various components of the electron transport chain. The linkage of redox states in electron transport systems to two component systems for gene regulation purposes is an attractive hypothesis (ALLEN 1993). It has been shown that in *Calothrix* strains differentiation into hormogonia and heterocysts is influenced by the redox state of the plastoquinone pool (CAMPBELL et al. 1993).

4 Combined stressors

Most work dealing with the effects of environmental loads on cyanobacteria, study only the effects of single stressors under otherwise optimum culturing conditions. In the natural environment, of course, various stressors usually act simultaneously, but work dealing with the effects of such situations on cyanobacteria are very rare. The first effect of almost all stressors is to inhibit growth. This general metabolic inhibition almost certainly triggers a general stress response accompanied by responses that are specific to each of the stressors. Evidence of such partial overlapping of various stress responses has been found in the case of *E. coli*. Studies into the synthesis of stress proteins suggests that this also applies to cyanobacteria. Various stressors such as UV radiation, oxidative stress, heavy metals, salt and darkness have been reported as enhancing the synthesis of typical heat shock proteins as well as their specific stress proteins (NICHOLSON et al. 1987, BHAGWAT and APTE 1989, BORBELY et al. 1990, SHIBATA et al. 1991, HAGEMANN et al. 1991, MITTLER and TEL-OR 1991).

Systematic studies of stress protein synthesis are a relatively simple method for identifying stress response groups and developing a regulone concept for cyanobacteria.

5 Future prospects

During the past few years, great progress has been made towards elucidating key biological processes such as photosynthesis and nitrogen-fixation in cyanobacteria. Numerous proteins and genes have been isolated from a wide range of strains and the functions described (TANDEAU de MARSAC and HOUMARD 1993). However, little is known about how these processes perform under the suboptimal growth conditions typical of the natural cyanobacterial habitat.

In the case of cyanobacteria, stress-related phenomena have been studied quite extensively for salt loading. Our knowledge of the mechanisms protecting against this stressor, such as the synthesis of low-molecular stress metabolites and high-molecular stress proteins, and of the metabolic adaptation process is quite large. The changes induced specifically by salt stress include active transport processes across the plasma membrane serving mainly the extrusion of surplus Na^+ ions. The prevailing opinion that this function is performed by a proton driven antiporter is being increasingly queried in favour of light induced primary extrusion by means of an Na^+ ATPase. Future studies must show whether the energetic basis of the plasma membrane is an Na^+ motive force and not the electrochemical proton gradient.

The few studies so far published on how stressors affect nitrogen metabolism have concentrated on quantitative rather than qualitative effects: e.g. the synthesis of nitrogenous osmoprotective substances accounting for up to 10% of the dry weight in response to salt stress binds considerable amounts of nitrogen; cells can lose up to 20% of their soluble protein by the secretion of drought stress proteins and their liberation during dehydration; a substantial nitrogen reservoir can be trapped by the remetabolisation of phycobiliproteins, which account for up to 50% of the soluble protein in a cell, during N starvation stress.

Investigations into the qualitative aspects of stress induced changes in nitrogen metabolism have dealt mainly with the extreme sensitivity of nitrogenase to salt and oxidative stress and variations induced by different nitrogen sources. Such studies must take dynamic aspects into account since factors and processes playing a hitherto unknown role may have substantial effects under the fluctuating and pessimal regime of the natural environment. This is shown by the example of the *NifJ* gene coding for a pyruvate flavodoxin oxidoreductase in the case of nitrogen-fixation. Under normal diazotrophic conditions, this gene is dispensable to *Anabaena* sp. PCC 7120, but under iron-deficient conditions it is essential for nitrogen-fixation (BAUER et al. 1993).

Our present knowledge is inadequate for modeling basic strategies for adaptation to abiotic stressors. As yet, few processes have been fully elucidated from the physiological down to the genetic level. The induction and characterization of mutants appears to be a suitable approach for such work. It will first be necessary to elucidate the nature and functions of the various stress proteins induced by a specific stressor or universally in response to various stress situations. Isolation, partial sequencing and comparison with the many known stress proteins should yield information about their function and enable the genes concerned to be cloned. The use of cloned genes to produce mutants in which the genes coding for the various stress proteins have been deleted is a relatively simple matter nowadays, and this in turn will reveal the roles of those genes and the products of their expression in the adaptation process.

Research into the interactions of stress situations should also yield interesting results. There is ample proof for "cross-protection" in heterotrophic eubacteria and yeasts, i.e. adaptation to one unfavourable parameter confers enhanced resistance to other, sometimes even lethal, stressors. For instance, after accumulating osmoprotective substances and salt stress proteins, salt adapted bacteria and yeasts exhibit enhanced heat tolerance (VÖLKER et al. 1992, PIPER 1993). Since the same stress proteins are sometimes induced by different stressors in cyanobacteria, there is some justification for expecting cross protection to occur in this group, too. Unavoidably, interest in the future will focus increasingly on regulatory mechanisms, i.e. the way in which external and internal signals are recognized and transformed into cellular responses.

Apart from being extraordinarily interesting basic research objects owing to their great evolutionary age, great adaptability and general usefulness as models of eukaryotic organisms, cyanobacteria are increasingly becoming the focus of biotechnological research. The spectrum of potential cyanobacterial products is large, ranging from toxins through pigments to biofertilizers to fix nitrogen (BOUSSIBA 1991, UTKILEN 1992, KERBY and ROWELL 1992). Cyanobacteria will be cultured under optimal growth conditions in closed systems to produce high-value chemicals needed in small quantities. In such systems, environmentally regulated promoters could be used systematically during certain growth phases to stimulate the expression of the desired product. In open systems designed for the production of biomass, and also when used as biofertilizers, cyanobacteria are exposed to varying conditions and a wide range of environmental stresses (RICHMOND 1992). Knowledge of the strategies that have evolved to cope with environmental loads and the selection and optimization of suitable strains for production will lead to higher yields. The composition of the biomass, the proportion of low molecular carbohydrates and the fatty acid composition can already be influenced fairly easily by choosing the optimal salinities and temperatures for the desired purpose.

However, further basic and applied research will be necessary to broaden the spectrum of biotechnical applications.

References

Aiba H, Nagaya M, Mizuno T (1993) Sensor and regulator proteins from the cyanobacterium *Synechococcus* species PCC 7942 that belong to the bacterial signal-transduction protein families: implication in the adaptive response to phosphate limitation. Molec Microbiol 8:81-91

Alge D, Peschek GA (1993) Characterization of a cta/CDE operon-like genomic region encoding subunits I-III of the cytochrome oxidase of the cyanobacterium *Synechocystis* PCC 6803. Biochem Molec Biol Int 29:511-525

Allen JF (1993) Redox control of transcription: sensors, response regulators, activators and repressors. FEBS Lett 332:203-207

Allen MM (1968) Photosynthetic membrane system in *Anacystis nidulans*. J Bacteriol 96:836-841

Apte SK (1992) Molecular biology of cyanobacterial nitrogen fixation: Recent advances. Indian J Microbiol 32:103-126

Apte SK, Bhagwat AA (1989) Salinity-stress-induced proteins in two nitrogen-fixing *Anabaena* strains differentially tolerant to salt. J Bacteriol 171:909-915

Apte SK, Haselkorn R (1990) Cloning of salinity stress-induced genes from the salt-tolerant nitrogen-fixing cyanobacterium *Anabaena torulosa*. Plant Molec Biol 15:723-733

Apte SK, Reddy BR, Thomas J (1987) Relationship between sodium influx and salt tolerance of nitrogen-fixing cyanobacteria. Appl Environ Microbiol 53:1934-1939

Apte SK, Thomas J (1986) Membrane electrogenesis and sodium transport in filamentous nitrogen-fixing cyanobacteria. Eur J Biochem 154:395-401

Batterton JC, van Baalen C (1971) Growth responses of blue-green algae to sodium chloride concentration. Arch Microbiol 76:151-165

Bauer CC, Scappino L, Haselkorn R (1993) Growth of the cyanobacterium *Anabaena* on molecular nitrogen: NifJ is required when iron is limited. Proc Natl Acad Sci USA 90:8812-8816

Belknap WR, Haselkorn R (1987) Cloning and light regulation of expression of the phycocyanin operon of the cyanobacterium *Anabaena*. EMBO J 6:871-884

Bhagwat AA, Apte SK (1989) Comparative analysis of proteins induced by heat shock, salinity, and osmotic stress in the nitrogen-fixing cyanobacterium *Anabaena* sp. strain L- 31. J Bacteriol 171:5187-51

Biggins J, Bruce D (1989) Regulation of excitation energy transfer in organisms containing phycobilins. Photosynth Res 20:1-34

Blondin PA, Kirby RJ, Barnum SR (1993) The heat shock response and acquired thermotolerance in three strains of cyanobacteria. Curr Microbiol 26:79-84

Bloye SA, Silman NJ, Mann NH, Carr NG (1992) Bicarbonate concentration by *Synechocystis* PCC 6803. Plant Physiol 99:601-606

Blumwald E, Mehlhorn RJ, Packer L (1983) Ionic osmoregulation during salt adaptation of the cyanobacterium *Synechococcus* 6311. Plant Physiol 73:377-380

Blumwald E, Tel-Or E (1982) Osmoregulation and cell composition in salt-adaptation of *Nostoc muscorum*. Arch Microbiol 132:168-172

Blumwald E, Tel-Or E. (1984) Salt adaptation of the cyanobacterium *Synechococcus* 6311 growing in a continuous culture (turbidostat). Plant Physiol 74:183-185

Blumwald E, Wolosin JM, Packer L (1984) Na^+/H^+ exchange in the cyanobacterium *Synechococcus* 6311. Biochem Biophys 122:452-459

Borbely G, Suranyi G, Korcz A, Palfi Z (1985) Effect of heat shock on protein synthesis in the cyanobacterium *Synechococcus* sp. strain PCC 6301. J Bacteriol 161:1125-1130

Borbely G, Suranyi G, Kos P (1990) Stress responses of cyanobacteria and the pleiotropic effects of light deprivation. FEMS Microbiol Ecol 74:141-152

Borowitzka LJ, Demmerle S, Mackay MA, Norton RS (1980) Carbon-13 nuclear magnetic resonance study of osmoregulation in a blue green alga. Science 210:650-651

Botsford JL, Harman JG (1992) Cyclic AMP in prokaryotes. Microbiol Rev 56:100-122

Boussiba S (1991) Nitrogen-fixing cyanobacteria potential uses. Plant Soil 137:177-180

Brahamsha B (1993) Identification of multiple RNA polymerase sigma factor homologs in cyanobacteria: Group 2 sigmas in *Anabaena* sp. strain PCC7120. Abstract, Cyanobacterial Workshop, Asilomar Conference Center Pacific Grove, CA

Brahamsha B, Haselkorn R (1991) Isolation and characterization of the gene encoding the principal sigma factor of the vegetative cell RNA polymerase from the cyanobacterium *Anabaena* sp. strain PCC 7120. J Bacteriol 173:2442-2450

Brahamsha B, Haselkorn R (1992) Identification of multiple RNA polymerase sigma factor homologs in the cyanobacterium *Anabaena* sp. strain PCC 7120: Cloning, expression, and inactivation of the sigB and sigC genes. J Bacteriol 174:7273-7282

Brown AD (1976) Microbial water stress. Bacteriol Rev 40:803-846

Brown II, Fadeyev SI, Gerasimenko LM, Kirik II, Pushenko MYA, Severina II (1990) Sodium ions are necessary for growth and energy transduction in the marine cyanobacterium *Oscillatoria brevis*. Arch Microbiol 153:409-411

Brunner G, Krauss H (1988) Deaggregation and proteolytic modification of a galactosyltransferase of *Poterioochromonas malhamensis*. Physiol Plant 74:708-714

Bryant DA (1982) Phycoerythrocyanin and phycoerythrin: Properties and occurence in cyanobacteria. J Gen Microbiol 128:835-844

Buikema WJ, Haselkorn R (1993) Molecular genetics of cyanobacterial development. Annu Rev Physiol 44:33-52

Bullerjahn GS, Riethman HC, Sherman LA (1987) Isolation and immunological characterization of a carotenoid-binding protein from the cyanobacteria *Synechocystis* sp. PCC 6714 and *Anacystis nidulans* R2. In: Biggins J (ed) Photosyn Res II. Martinus Nijhoff, Dordrecht, pp 145-148

Bullerjahn GS, Sherman LA (1986) Identification of a carotenoid-binding protein in the cytoplasm membrane from the heterotrophic cyanobacterium *Synechocystis* strain PCC 6714. J Bacteriol 167:396-399

Campbell D, Houmard J, Tandeau de Marsac N (1993) Electron transport regulates cellular differentiation in the filamentous cyanobacterium *Calothrix*. Plant Cell 5:451-463

Chitnis PR, Nelson N (1991) Molecular cloning of the genes encoding two chaperone proteins of the cyanobacterium *Synechocystis* sp. PCC 6803. J Biol Chem 266:58-65

Close TJ, Lammers PJ (1993) An osmotic stress protein of cyanobacteria is immunologically related to plant dehydrins. Plant Physiol 101:773-779

Cookson MJ, Baird PN, Hall LMC, Coates ARM (1989) Identification of two unknown reading frames in *Synechococcus* 6301 as homologues of the 10K and 65K antigen genes of *Mycobacterium tuberculosis* and related heat shock genes in *E. coli* and *Coxiella burnettii*. Nucleic Acids Res 17:6392

Crawford NA, Yee BC, Droux M, Carlson DE, Buchanan BB (1988) Ferredoxin/Thioredoxin system. In: Packer L, Glazer G (eds) Methods in enzymology, vol 167. Academic Press, San Diego, pp 415-427

Cseke C, Balogh A, Farkas GL (1981) Redox modulation of glucose 6-phosphate dehydrogenase in *Anacystis nidulans* and its "uncoupling" by phage infection. FEBS Lett 126:85-88

Csonka LN, Hanson AD (1991) Prokaryotic osmoregulation: Genetics and physiology. Annu Rev Microbiol 45:569-606

de Lorimier RM, Smith RL, Stevens SE Jr (1992) Regulation of phycobilisome structure and gene expression by light intensity. Plant Physiol 98:1003-1010

Dewar MA, Barber J (1974) Chloride transport in *Anacystis nidulans*. Planta 117:163-172

Dubois JD, Kapustka LA (1981) Osmotic stress effects on the N_2 (C_2H_2)ase activity of aquatic cyanobacteria. Aquatic Bot 11:11-20

Erdmann N (1983) Organic osmoregulatory solutes in blue-green algae. Z Pflanzenphys 110:147-155

Erdmann N, Berg C, Hagemann M (1989) Missing salt adaptation of *Microcystis firma* (cyanobacterium) in the dark. Arch Hydrobiol 114:521-530

Erdmann N, Fulda S, Hagemann M (1992) Glucosylglycerol accumulation during salt acclimation of two unicellular cyanobacteria. J Gen Microbiol 138:363-368

Erdmann N, Schiewer U (1985) $^{14}CO_2$ fixation pattern of cyanobacteria. Biochem Physiol Pflanzen 180:515-532

Espie GS, Miller AG, Canvin DT (1988) Characterization of the Na^+-requirement in cyanobacterial photosynthesis. Plant Physiol 88:757-763

Faulkner M (1988) Polypeptide patterns obtained by polyacrylamide gel electrophoresis of thylakoid membranes isolated from light and dark grown strains of *Chlorogloea fritschii*. Electrophoresis 9:807-811

Fernandes TA, Iyer V, Apte SK (1993) Differential response of nitrogen-fixing cyanobacteria to salinity and osmotic stresses. Appl Environ Microbiol 59:899-904

Fernández VE, Avendaño MC (1993) Sodium- stimulation of phosphate uptake in the cyanobacterium *Anabaena* PCC 7119. Plant Cell Physiol 34:201-207

Fogg GE, Stewart WDP, Fay P, Walsby AE (eds) (1973) The blue-green algae. Academic Press, London and New York

Fork DC, Herbert SK (1993) Electron transport and photophosphorylation by Photosystem I *in vivo* in plants and cyanobacteria. Photosyn Res 36:149-168

Fujita Y, Murakami A, Ohki K (1990) Regulation of stoichiometry of thylakoid components in the photosynthetic system of cyanophytes: Model experiments showing that control of the synthesis or supply of chl *a* can change the stoichiometric relationship between the two photosystems. Plant Cell Physiol 31:145-153

Fujita Y, Murakami A, Ohki K, Hagiwara N (1988) Regulation of photosystem composition in cyanobacterial photosynthetic system: Evidence indicating that Photosystem I formation is controlled in response to the electron transport state. Plant Cell Physiol 29:557-564

Fulda S, Hagemann M, Libbert E (1990) Release of glucosylglycerol from the cyanobacterium *Synechocystis* spec. SAG 92.79 by hypoosmotic shock. Arch Microbiol 153:405-408

Gabbay-Azaria R, Pick U, Ben-Hayyim G, Tel-Or E (1994) The involvement of a vanadate-sensitive ATPase in plasma membranes of a salt tolerant cyanobacterium. Physiol Plant 90:692-698

Gabbay-Azaria R, Schonfeld M, Tel-Or E (1989) Cytochrome oxidase and H^+-ATPase activities in plasma-membranes of the marine cyanobacterium *Spirulina subsala* and their possible role in salt tolerance. In: Dainty J et al. (eds) Plant Membrane Transport. Elsevier Sci Publ, pp 687-688

Gabbay-Azaria R, Schonfeld M, Tel-Or S, Messinger R, Tel-Or E (1992) Respiratory activity in the marine cyanobacterium *Spirulina subsalsa* and its role in salt tolerance. Arch Microbiol 157:183-190

Gabbay-Azaria R, Tel-Or E (1993) Mechanisms of salt tolerance in cyanobacteria. In: Gresshoff PM (ed) Plant responses to the environment. CRC Press, Boca Raton, pp 123-132

Gabbay-Azaria R, Tel-Or E, Schonfeld M (1988) Glycinebetaine as an osmoregulant and compatible solute in the marine cyanobacterium *Spirulina subsalsa*. Arch Biochem Biophys 261:333-339

Giaever HM, Styrvold OB, Kaasen I, Strom AR (1988) Biochemical and genetic characterization of osmoregulatory trehalose synthesis in *Escherichia coli*. J Bacteriol 170:2841-2849

Goldstein J, Pollitt NS, Inouye M (1990) Major cold shock protein of *Escherichia coli*. Proc Natl Acad Sci USA 87:283-287

Goloubinoff P, Christeller JT, Gatenby AA, Lorimer GH (1989a) Reconstitution of active dimeric ribulose bisphosphate carboxylase from an unfolded state depends on two chaperonin proteins and Mg-ATP. Nature 342:884-888

Goloubinoff P, Gatenby AA, Lorimer GH (1989b) *GroE* heat- shock proteins promote assembly of foreign prokaryotic ribulose bisphosphate carboxylase oligomers in *Escherichia coli*. Nature 337:44-47

Gombos Z, Vigh L (1986) Primary role of the cytoplasmic membrane in thermal acclimation evidenced in nitrate-starved cells of the blue- green alga, *Anacystis nidulans*. Plant Physiol 80:415-419

Gombos Z, Wada H, Hideg E, Murata N (1994) The unsaturation of membrane lipids stabilizes photosynthesis against heat stress. Plant Physiol 104:563-567

Gombos Z, Wada H, Murata N (1991) Direct evaluation of effects of fatty-acid unsaturation on the thermal properties of photosynthetic activities, as studied by mutation and transformation of *Synechocystis* PCC 6803. Plant Cell Physiol 32:205-211

Grilli Caiola M, Canini A, Galiazzo, F, Rotilio, G (1991) Superoxide dismutase in vegetative cells, heterocysts and akinetes of *Anabaena cylindrica* Lemm. FEMS Microbiol Lett 80:161-166

Grossman AR, Schaefer MR, Chiang GG, Collier JL (1993) The phycobilisome, a light-harvesting complex responsive to environmental conditions. Microbiol Rev 57:725-749

Gruber MY, Glick BR, Thompson JE (1990) Cloned manganese superoxide dismutase reduces oxidative stress in *Escherichia coli* and *Anacystis nidulans*. Proc Natl Acad Sci USA 87:2608-2612

Hagemann M, Erdmann N (1994) Activation and pathway of glucosyl glycerol synthesis in the cyanobacterium *Synechocystis* sp. PCC 6803. Microbiol 140:1427-1431

Hagemann M, Erdmann N Wittenburg E (1989a) Studies concerning enzyme activities in salt-loaded cells of the cyanobacterium *Microcystis firma*. Biochem Physiol Pflanzen 184:87-94

Hagemann M, Erdmann N, Schiewer U (1989b) Salt adaptation of the cyanobacteria *Microcystis firma* and *Synechocystis aquatilis* in turbidostat cultures. I. Steady state values. Arch Hydrobiol Suppl 82:425-435

Hagemann M, Erdmann N, Wittenburg E (1987) Synthesis of glucosylglycerol in salt-stressed cells of the cyanobacterium *Microcystis firma*. Arch Microbiol 148:275-279

Hagemann M, Fulda S, Schubert H (1994) DNA, RNA, and protein synthesis in the cyanobacterium *Synechocystis* sp. PCC 6803 adapted to different salt concentrations. Curr Microbiol 28:201-207

Hagemann M, Golldack, D, Biggins J, Erdmann N (1993) Salt-dependent protein phosphorylation in the cyanobacterium *Synechocystis* PCC 6803. FEMS Microbiol Lett 113:205-210

Hagemann M, Techel D, Rensing L (1991) Comparison of salt- and heat-induced alterations of protein synthesis in the cyanobacterium *Synechocystis* sp. PCC 6803. Arch Microbiol 155:587-592

Hagemann M, Wölfel L, Krüger B (1990) Alterations of protein synthesis in the cyanobacterium *Synechocystis* sp. PCC 6803 after a salt shock. J Gen Microbiol 136:1393-1399

Hagemann M, Zuther E (1992) Selection and characterization of mutants of the cyanobacterium *Synechocystis* sp. PCC 6803 unable to tolerate high salt concentrations. Arch Microbiol 158:429-434

Hansel A, Schmid A, Tadros MH, Jürgens UJ (1994) Isolation and characterisation of porin from the outer membrane of *Synechococcus* PCC 6301. Arch Microbiol 161:163-167

Hanson AD, May AM, Grumet R, Bode J, Jamieson GC, Rhodes D (1985) Betaine synthesis in chenopods: Localization in chloroplasts. Proc Natl Acad Sci USA 82:3678-3682

Harrison MA, Keen JN, Findlay JBC, Allen JF (1990) Modification of a glnB-like gene product by photosynthetic electron transport in the cyanobacterium *Synechococcus* 6301. FEBS Lett 264:25-28

Hershkovitz N, Oren A, Cohen Y (1991a) Accumulation of trehalose and sucrose in cyanobacteria exposed to matric water stress. Appl Environ Microbiol 57:645-648

Hershkovitz N, Oren A, Post A, Cohen Y (1991b) Induction of water-stress proteins in cyanobacteria exposed to matric- or osmotic-water stress. FEMS Microbiol Lett 83:169-172

Hill DR, Hladun SL, Scherer S, Potts M (1994) Water stress proteins of *Nostoc commune* (Cyanobacteria) are secreted with UV-A/B- absorbing pigments and associate with 1,4-β-D-xylanxylano-hydrolase activity. J Biol Chem 269:7726-7734

Huber SC, Huber JL (1992) Role of sucrose-phosphate synthase in sucrose metabolism in leaves. Plant Physiol 99:1275-1278

Incharoensakdi A, Takabe T Akazawa T (1986) Effect of betaine on enzyme activity and subunit interaction of ribulose-1,5-bisphosphate carboxylase/oxygenase from *Aphanothece halophytica.* Plant Physiol 81:1044-1049

Incharoensakdi A, Takabe T (1988) Determination of intracellular chloride ion concentration in a halotolerant cyanobacterium *Aphanothece halophytica.* Plant Cell Physiol 29:1073-1075

Ishitani M, Takabe T, Kojima K, Takabe T (1993) Regulation of glycinebetaine accumulation in the halotolerant cyanobacterium *Aphanothece halophytica.* Aust J Plant Physiol 20:693-703

Jäger KM, Bergman B (1990) Localization of a multifunctional chaperonin (GroEL protein) in nitrogen-fixing *Anabaena* PCC 7120. Planta 183:120-125

Jeanjean R, Matthijs HCP, Onana B, Havaux M, Joset F (1993) Exposure of the cyanobacterium *Synechocystis* PCC 6803 to salt stress induces concerted changes in respiration and photosynthe sis. Plant Cell Physiol 34:1073-1079

Jeanjean R, Onana B, Peschek GA, Joset F (1990) Mutants of the cyanobacterium *Synechocystis* PCC 6803 impaired in respiration and unable to tolerate high salt concentrations. FEMS Microbiol Lett 68:125-130

Jones PG, van Bogelen RA, Neidhardt FC (1987) Induction of proteins in response to low temperature in *Escherichia coli*. J Bacteriol 169:2092-2095

Kanamaru K, Kashiwagi S, Mizuno T (1993) The cyanobacterium *Synechococcus* sp PCC 7942, possesses two distinct genes encoding cation-transporting P-type ATPases. FEBS Lett 330:99-104

Kanervo E, Mäenpää P, Aro E-M (1993) D1 protein degradation and psbA transcript levels in *Synechocystis* PCC 6803 during photoinhibition *in vivo*. J Plant Physiol 142:669-675

Kauss H (1987) Some experiments of calcium-dependent regulation in plant metabolism. Annu Rev Plant Physiol 38:47-72

Kerby NW, Reed RH, Rowell P (1990) Incorporation of ^{14}C in the cyanobacterium *Synechococcus* PCC 6301 following salt stress. Planta 181:393-398

Kerby NW, Rowell P (1992) Potential and commercial applications for photosynthetic prokaryotes. In: Mann NH, Carr NG (eds) Photosynthetic Prokaryotes, Biotechnology Handbooks, vol 6. Plenum Press, New York, pp 233-265

Khanna R, Graham JR, Myers J, Gantt E (1983) Phycobilisome composition and possible relationship to reaction centers. Arch Biochem Biophys 224:534-542

Khomutov G, Fry IV, Huflejt ME, Packer L (1990) Membrane lipid composition, fluidity, and surface charge changes in response to growth of the freshwater cyanobacterium *Synechococcus* 6311 under high salinity. Arch Biochem Biophys 277:263-267

Kogut, M (1991) The 'true' intracellular environment of moderately halophilic eubacteria. In: Rodriguez-Valera F (ed) General and applied aspects of halophilic microorganisms. Plenum Press, New York, pp 217-224

Kollman VH, Hanners LJ, London RE, Adame EG, Walker TE (1979) Photosynthetic preparation and characterization of ^{13}C-labeled carbohydrates in *Agmenellum quadruplicatum*. Carbohydr Res 73:193-202

König F (1990) Shade adaptation in cyanobacteria. Further characterization of *Anacystis* shade phenotype as induced by sublethal concentrations of DCMU-type inhibitors in strong light. Photosynth Res 26:29-37

Koo S-P, Higgins CF, Booth IR (1991) Regulation of compatible solute accumulation in *Salmonella typhimurium*: evidence for glycine betaine efflux system. J Gen Microbiol 137:2617-2625

Krupa Z, Öquist G, Gustafsson P (1990) Photoinhibition and recovery of photosynthesis in psbA gene-inactivated strains of cyanobacterium *Anacystis nidulans*. Plant Physiol 93:1-6

Kuschner DJ (1985) The halobacteriaceae. In: Woese CR, Wolfe RS (eds) Bacteria: a treatise on structure and function, Archaebacteria, vol 8. Academic Press, Orlando, pp 171-184

Laudenbach DE, Trick CG, Strauss NA (1989) Cloning and characterization of an *Anacystis nidulans* R2 superoxide dismutase gene. Mol Gen Genet 216:455-461

Lefort Tran M, Pouphile M, Spath S (1988) Cytoplasmic membrane changes during adaptation of the freshwater cyanobacterium *Synechococcus* 6311 to salinity. Plant Physiol 87:767-775

Lehel C, Gombos Z, Török Z, Vigh L (1993) Growth temperature modulates thermotolerance and heat shock response of cyanobacterium *Synechocystis* PCC 6803. Plant Physiol Biochem 31:81-88

Lehel C, Wada H, Kovacs E, Török Z, Gombos Z, Horvath I, Murata N, Vigh L (1992) Heat shock protein synthesis of the cyanobacterium *Synechocystis* PCC 6803: Purification of the GroEL-related chaperonin. Plant Molec Biol 18:327-336

Mackay MA, Norton RS (1987) ^{13}C nuclear magnetic resonance study of biosynthesis of glucosylglycerol by a cyanobacterium under osmotic stress. J Gen Microbiol 133:1535-1542

Mackay MA, Norton RS, Borowitzka LJ (1983) Marine blue-green algae have a unique osmoregulatory system. Mar Biol 73:301-307

Mackay MA, Norton RS, Borowitzka LJ (1984) Organic osmoregulatory solutes in cyanobacteria. J Gen Microbiol 130:2177-2191

Mamedov, MD, Hayashi H, Wada H, Mohanty PS, Papageorgiou GC, Murata N (1991) Glycinebetaine enhances and stabilizes the evolution of oxygen and the synthesis of ATP by cyanobacterial thylakoid membranes. FEBS Lett 294:271-274

Mann NH, Carr NG, Midgley JEM (1975) RNA synthesis and the accumulation of guanine nucleotides during growth shift down in the blue-green alga *Anacystis nidulans*. Biochim Biophys Acta 402:41-50

Mann NH, Rippka R, Herdman M (1991) Regulation of protein phosphorylation in the cyanobacterium *Anabaena strain* PCC 7120. J Gen Microbiol 137:331-339

Mannan RM, Pakrasi HB (1993) Dark heterotrophic growth conditions result in an increase in the content of Photosystem II units in the filamentous cyanobacterium *Anabaena variabilis* ATCC 29413. Plant Physiol 103:971-977

Marino GT, Asato Y (1986) Characterization of cell cycle events in the dark in *Anacystis nidulans*. J Gen Microbiol 132:2123-2127

Martel A, Yu S, Garcia-Reina G, Lindblad P, Pedersén M (1992) Osmotic-adjustment in the cyanobacterium *Spirulina platensis*: Presence of an α-glucosidase. Plant Physiol Biochem 30:573-578

Martin J, Horwich AL, Hartl FU (1992) Prevention of protein denaturation under heat stress by the chaperonin Hsp 60. Science 258:995-998

Masamoto K, Riethman HC, Sherman LA (1987) Isolation and characterization of a carotenoid-associated thylakoid protein from the cyanobacterium *Anacystis nidulans* R2. Plant Physiol 84:633-639

Matthijs HCP, Lubberding HJ (1988) Dark respiration in cyanobacteria. In: Rogers LJ, Gallon JR (eds) Biochemistry of the algae and cyanobacteria. Clarendon Press, Oxford, pp 131-145

Mehlhorn RJ, Sullivan R (1988) Measurement of bioenergetics phenomena in cyanobacteria with magnetic resonance techniques. In: Packer L, Glazer AN (eds) Methods in enzymology, vol 167. Academic Press, San Diego, pp 509-518

Miller DM, Jonas JH, Yopp JH, Tindall DR, Schmid WE (1976) Ion metabolism in a halophilic blue-green alga, *Aphanothece halophytica*. Arch Microbiol 111:145-149

Miller M, Cox RP, Simpson D (1989) Effect of growth temperature on the composition of the photosynthetic apparatus in a thermophilic cyanobacterium (*Synechococcus* sp.). FEMS Microbiol Lett 57:39-44

Mittler R, Tel-Or E (1991) Oxidative stress responses and shock proteins in the unicellular cyanobacterium *Synechococcus* R2 (PCC-7942). Arch Microbiol 155:125-130

Mohammad FAA, Reed RH, Stewart WDP (1983) The halophilic cyanobacterium *Synechocystis* DUN 52 and its osmotic responses. FEMS Microbiol Lett 16:287-290

Molitor V, Erber W, Peschek GA (1986) Increased levels of cytochrome oxidase and sodium-proton antiporter in the plasma membrane of *Anacystis nidulans* after growth in sodium-enriched media. FEBS Lett 204:251-256

Molitor V, Kunter O, Sleytr UB, Peschek GA (1990a) The ultrastructure of plasma and thylakoid membrane vesicles from the fresh water cyanobacterium *Anacystis nidulans* adapted to salinity. Protoplasma 157:112-119

Molitor V, Trnka M, Erber W, Steffan I, Riviere M-E, Arrio B, Springer-Lederer H, Peschek GA (1990b) Impact of salt adaptation on esterified fatty acids and cytochrome oxidase in plasma and thylakoid membranes from the cyanobacterium *Anacystis nidulans*. Arch Microbiol 154:112-119

Moore DJ, Reed RH, Stewart WDP (1985) Responses of cyanobacteria to low level osmotic stress: Implications for the use of buffers. J Gen Microbiol 131:1267-1272

Moore DJ, Reed RH, Stewart WDP (1987) A glycine betaine transport system in *Aphanothece halophytica* and other glycine betaine-synthesising cyanobacteria. Arch Microbiol 147:399-405

Murakami A, Fujita Y (1991) Regulation of photosystem stoichiometry in the photosynthetic system of the cyanophyte *Synechocystis* PCC 6714 in response to light-intensity. Plant Cell Physiol 32:223-230

Murata N (1989) Low-temperature effects on cyanobacterial membranes. J Bioenerg Biomem 21:61-75

Neidhardt FC, Van Bogelen RA (1987) Heat Shock response. In: Neidhardt et al. (eds) *Escherichia coli* and *Salmonella typhimurium*: cellular and molecular biology. American Soc Microbiol, Washington, pp 1334-1345

Nicholson P, Osborn RW, Howe CJ (1987) Induction of protein synthesis in response to ultraviolet light, nalidixic acid and heat shock in the cyanobacterium *Phormidium laminosum*. FEBS Lett 221:110-114

Nicholson P, Varley JPA, Howe CJ (1991) A comparison of stress responses in the cyanobacterium *Phormidium laminosum*. FEMS Microbiol Lett 78:109-114

Nitschmann WH, Packer L (1992) NMR studies on Na^+ transport in *Synechococcus* PCC 6311. Arch Biochem Biophys 294:347-352

Nitschmann, WH, Peschek GA (1985) Modes of proton translocation across the cell membrane of respiring cyanobacteria. Arch Microbiol 141:330-336

Niven GW, Kerby NW, Rowell P, Reed RH, Stewart WDP (1987) The effects of salt on nitrogen-fixation and ammonium assimilation in *Anabaena variabilis*. Br Phycol J 22:411-416

Nomura M, Ishitani M, Takabe T, Rai AK, Takabe T (1995) *Synechococcus* sp. PCC 7942 transformed with *Escherichia coli bet* genes produces glycine betaine from choline and acquires resistance to salt stress. Plant Physiol 107:703-708

Ohmori K, Hirose M, Ohmori M (1992) Function of cAMP as a mat-forming factor in the cyanobacterium *Spirulina platensis*. Plant Cell Physiol 33:21-25

Ohmori M (1989) cAMP in *Anabaena cylindrica*: Rapid changes in Cellular levels in response to changes in extracellular environments. Plant Cell Physiol 30:911-914

Ohmori M, Ohmori K, Hasunuma K (1988) Rapid change in cyclic 3',5'-AMP concentration triggered by a light-off or light-on signal in *Anabaena cylindrica*. Arch Microbiol 150:203-204

Onana B, Jeanjean R, Joset F (1994) A gene, *stpA*, involved in the establishment of salt tolerance in the cyanobacterium *Synechocystis* PCC 6803. Russian J Plant Physiol 41:152-159

Padan E, Schuldiner S (1993) Na^+/H^+antiporters, molecular devices that couple the Na^+ and H^+ circulation in cells. J Bioenerg Biomem 25:647-669

Padan E, Vitterbo A (1988) Cation transport in cyanobacteria. In: Packer L, Glazer AN (eds) Methods in enzymology, vol 167. Academic Press, San Diego, pp 561-571

Palfi Z, Suranyi G, Borbely G (1991) Alterations in the accumulation of adenylylated nucleotides in heavy-metal-ion- stresses and heat-stressed *Synechococcus* sp. strain PCC 6301, a cyanobacterium, in light and dark. Biochem J 276:487-491

Paleg LG, Stewart GR, Starr R (1985) The effect of compatible solutes on proteins. Plant Soil 89:83-94

Parsell DA, Lindquist S (1993) The function of heat-shock proteins in stress tolerance: Degradation and reactivation of damaged proteins. Annu Rev Genet 27:437-496

Paschinger H (1977a) DCCD induced sodium uptake by *Anacystis nidulans*. Arch Microbiol 113:285-291

Paschinger H (1977b) Sodium chloride stimulated respiration of *Anacystis nidulans*. Z Allg Mikrobiol 17:373-379

Peat A, Powell N, Potts M (1988) Ultrastructural analysis of the rehydration of desiccated *Nostoc commune* HUN (cyanobacteria) with particular reference to the immunolabelling of NifH. Protoplasma 146:72-80

Peschek GA, Molitor V, Trnka M, Wastyn M, Erber W (1988) Characterization of cytochrome-c oxidase in isolated and purified plasma and thylakoid membranes from cyanobacteria. In: Packer L., Glazer AN (eds) Methods in enzymology, vol 167. Academic Press, San Diego, pp 437-449

Piper PW (1993) Molecular events associated with acquisition of heat tolerance by the yeast *Saccharomyces cerevisiae*. FEMS Microbiol Rev 11:339-356

Pollard A, Wyn Jones RG (1979) Enzyme activities in concentrated solutions of glycine betaine and other solutes. Planta 144:292-298

Potts M, Bowman MA, Morrison NS (1984) Control of matric water potential (ψ_m) in immobilised cultures of cyanobacteria. FEMS Microbiol Lett 24:193-196

Potts M, Ocampo-Friedmann R, Bowman MA, Tözun B (1983) *Chroococcus* N41 (cyanobacteria): morphological, biochemical and genetic characterization and effects of water stress on ultrastructure. Arch Microbiol 135:81-90

Potts M, Sun Hong, Mockaitis K, Kennelly PJ, Reed D, Tonks NK (1993) A protein-tyrosine/serine phosphatase encoded by the genome of the cyanobacterium *Nostoc commune* UTEX 584. J Biol Chem 268:7632-7635

Rady AA, Elsheekh MM, Matkovics B (1994) Temperature shift-induced changes in the antioxidant enzyme system of cyanobacterium *Synechocystis* PCC-6803. Int J Biochem 26:433-435

Rai AK (1990) Biochemical characteristics of photosynthetic response to various external salanities in halotolerant and freshwater cyanobacteria. FEMS Microbiol Lett 69:177-180

Rai AK, Abraham G (1993) Salinity tolerance and growth analysis of the cyanobacterium *Anabaena doliolum*. Bull Environ Contam Toxicol 51:724-731

Rai AK, Abraham G (1995) Relationship of combined nitrogen sources to salt tolerance in freshwater cyanobacterium *Anabaena doliolum*. J Appl Bacteriol 78:501-506

Raps S, Kycia JH, Ledbetter MC, Siegelman HW (1985) Light intensity adaptation and phycobilisome composition of *Microcystis aeruginosa*. Plant Physiol 79:983-987

Raps S, Wyman K, Siegelman HW, Falkowski PG (1983) Adaptation of the cyanobacterium *Microcystis aeruginosa* to light intensity. Plant Physiol 72:829-832

Raven JA (1985) Regulation of pH and generation of osmolarity in vascular plants: A cost-benefit analysis in relation to efficiency of use of energy, nitrogen and water. New Phytol 101:25-77

Reddy BR, Apte SK, Thomas J (1989) Enhancement of cyanobacterial salt tolerance by combined nitrogen. Plant Physiol 89:204-210

Reed RH (1984) Use and abuse of osmo-terminology. Plant Cell Environ 7:165-170

Reed RH, Borowitzka LJ, Mackay MA, Chudek JA, Foster R, Warr SRC, Moore DJ, Stewart WDP (1986a) Organic solute accumulation in osmotically stressed cyanobacteria. FEMS Microbiol Rev 39:51-56

Reed RH, Chudek JA, Foster R, Stewart WDP (1984a) Osmotic adjustment in cyanobacteria from hypersaline environments. Arch Microbiol 138:333-337

Reed RH, Richardson DL, Stewart WDP (1985a) Na^+ uptake and extrusion in the cyanobacterium *Synechocystis* PCC 6714 in response to hypersaline treatment. Evidence for transient changes in plasmalemma Na^+ permeability. Biochim Biophys Acta 814:347-355

Reed RH, Richardson DL, Warr SRC, Stewart WDP (1984b) Carbohydrate accumulation and osmotic stress in cyanobacteria. J Gen Microbiol 130:1-4

Reed RH, Rowell P, Stewart WDP (1981) Characterization of the transport of potassium ions in the cyanobacterium *Anabaena variabilis* Kütz. J Biochem 116:323-330

Reed RH, Stewart WDP (1983) Physiological responses of *Rivularia atra* to salinity: Osmotic adjustment in hyposaline media. New Phytol 95:595-603

Reed RH, Stewart WDP (1985) Osmotic adjustment and organic solute accumulation in unicellular cyanobacteria from freshwater and marine habitats. Mar Biol 88:1-9

Reed RH, Warr SRC, Kerby NW, Stewart WDP (1986b) Osmotic shock-induced release of low molecular weight metabolites from free-living and immobilized cyanobacteria. Enzyme Microb Technol 8:101-104

Reed RH, Warr SRC. Richardson DL, Moore DJ, Stewart WDP (1985b) Multiphasic osmotic adjustment in a euryhaline cyanobacterium. FEMS Microbiol Lett 28:225-229

Reuter W, Müller C (1993) Adaptation of the photosynthetic apparatus of cyanobacteria to light and CO_2. J Photochem Photobiol 21:3-27

Reuter W, Nickel-Reuter C (1993) Molecular assembly of the phycobilisomes from *Mastigocladus laminosus*. J Photochem Photobiol 18:51-66

Riccardi G, Cella R, Camerino G, Ciferri O (1983) Resistance to azetidine-2-carboxylic acid and sodium chloride tolerance in carrot cell cultures and *Spirulina platensis*. Plant Cell Physiol 24:1073-1078

Richmond A (1992) Mass culture of cyanobacteria. In: Mann NH, Carr NG (eds) Photosynthetic prokaryotes, Biotechnology handbook, vol 6. Plenum Press, New York, pp 181-210

Ritchie RJ (1992a) Sodium transport and the origin of the membrane potential in the cyanobacterium *Synechococcus* R-2 (*Anacystis nidulans*) PCC 7942. J Plant Physiol 139:320-330

Ritchie RJ (1992b) The cyanobacterium *Synechococcus* R-2 (*Anacystis nidulans, S. leopoliensis*) PCC 7942 has a sodium-dependent chloride transporter. Plant Cell Environ 15:163-177

Ritter D, Yopp JH (1993) Plasma membrane lipid composition of the halophilic cyanobacterium *Aphanothece halophytica*. Arch Microbiol 159:435-439

Rowell P, Darling AJ, Stewart WDP (1988) Thioredoxin system. In: Packer L, Glazer AN (eds) Methods in enzymology, vol 167. Academic Press, San Diego, pp 410-415

Russel NJ (1989) Adaptive modifications in membranes of halotolerant and halophilic microorganisms. J Bioenerg Biomem 21:93-113

Saier jr MH (1993) Introduction: Protein phosphorylation and signal transduction in bacteria. J Cell Biochem 51:1-6

Sakamoto T, Murata N, Ohmori M (1991) The concentration of cyclic AMP and adenylate cyclase activity in cyanobacteria. Plant Cell Physiol 32:581-584

Samuelsson G, Lönneborg A, Gustafsson P, Öquist G (1987) The susceptibility of photosynthesis to photoinhibition and the capacity of recovery in high and low light grown cyanobacterium *Anacystis nidulans*. Plant Physiol 83:438-441

Sanders CE, Melis A, Allen JF (1989) In vivo phosphorylation of proteins in the cyanobacterium *Synechococcus* 6301 after chromatic acclimation to Photosystem I or Photosystem II light. Biochim Biophys Acta 976:168-172

Santoro MM, Liu Y, Khan SMA, Hou L-X, Bolen DW (1992) Increased thermal stability of proteins in the presence of naturally occurring osmolytes. Biochem 31:5278-5283

Sato N (1992) Cloning of a low-temperature-induced gene lti2 from the cyanobacterium *Anabaena variabilis* M3 that is homologous to Ó-amylases. Plant Molec Biol 18:165-170

Sato N (1994) A cold-regulated cyanobacterial gene cluster encodes RNA-binding protein and ribosomal protein S 21. Plant Molec Biol 24:819-823

Sato N, Murata N (1980) Temperature shift-induced responses in lipids in the blue-green alga, *Anabaena variabilis*. The central role of diacylmonogalactosylglycerol in thermoadaptation. Biochim Biophys Acta 619:353-366

Schaefer MR, Golden SS (1989a) Light availability influences the ratio of two forms of D1 in cyanobacterial thylakoids. J Biol Chem 264:7412-7417

Schaefer MR, Golden SS (1989b) Differential expression of members of a cyanobacterial psbA gene family in response to light. J Bacteriol 171:3973-3981

Scherer S, Ernst A, Chen T-W, Böger P (1984) Rewetting of drought-resistant blue-green algae: Time course of water uptake and reappearance of respiration, photosynthesis, and nitrogen fixation. Oecologia 62:418-423

Scherer S, Potts M (1989) Novel water stress protein from a desiccation-tolerant cyanobacterium. Purification and partial characterization. J Biol Chem 264:12546-12553

Scherer S, Riege H, Böger P (1988) Light-induced proton efflux of the cyanobacterium *Anabaena variabilis*. In: Rogers LJ, Gallon JR (eds) Biochemistry of the algae and cyanobacteria. Clarendon Press, Oxford, pp 121-130

Schiewer U, Jonas I (1977) Die Wirkung unterschiedlicher NaCl-Konzentrationen auf die Ultrastruktur von Blaualgen. I. *Microcystis firma*. Arch Protistenk 119:127-145

Schobert B (1980) The importance of water activity and water structure during hyperosmotic stress in algae and higher plants. Biochem Physiol Pflanzen 175:91-103

Schubert H, Fulda S, Hagemann M (1993) Effects of adaptation to different salt concentrations on photosynthesis and pigmentation of the cyanobacterium *Synechocystis* sp PCC 6803. J Plant Physiol 142:291-295

Schubert H, Hagemann M (1990) Salt effects on 77K fluorescence and photosynthesis in the cyanobacterium *Synechocystis* sp. PCC 6803. FEMS Microbiol Lett 71:169-172

Shibata H, Baba K, Ochiai H (1991) Near-UV irradiation induces shock proteins in *Anacystis nidulans* R-2; possible role of active oxygen. Plant Cell Physiol 32:771-776

Shyam R, Raghavendra AS, Sane PV (1993) Role of dark respiration in photoinhibition of photosynthesis and its reactivation in the cyanobacterium *Anacystis nidulans*. Physiol Plant 88:446-452

Sibley MH, Yopp JH (1987) Regulation of S- adenosylhomocysteine hydrolase in the halophilic cyanobacterium *Aphanothece halophytica*: a possible role in glycinebtaine biosyntheses. Arch Microbiol 149:43-46

Sobczyk A, Schyns G, de Marsac NT, Houmard J (1993) Transduction of the light signal during complementary chromatic adaptation in the cyanobacterium *Calothrix* sp. PCC 7601: DNA-binding proteins and modulation by phosphorylation. EMBO J 12:997-1004

Stal LJ, Reed RH (1987) Low-molecular mass carbohydrate accumulation in cyanobacteria from a marine microbial mat in response to salt. FEMS Microbiol Ecol 45:305-312

Stanier RY, Cohen-Bazire G (1977) Phototrophic prokaryotes: The cyanobacteria. Annu Rev Microbiol 31:225-274

Stock JB, Ninfa AJ, Stock AM (1989) Protein phosphorylation and regulation of adaptive responses in bacteria. Microbiol Rev 53:450-490

Streb P, Michael-Knauf A, Feuerabend J (1993) Preferential photoinactivation of catalase and photoinhibition of Photosystem II are common early symptoms under various osmotic and chemical stress conditions. Physiol Plant 88:590-598

Strom AR, Falkenberg P, Landfald B (1986) Genetics of osmoregulation in *Escherichia coli*: uptake and biosynthesis of organic osmolytes. FEMS Microbiol Rev 39:79-86

Suranyi G, Korsz A, Palfi Z, Borbely G (1987) Effects of light deprivation on RNA synthesis, accumulation of Guanosine3'(2')-diphosphate 5'-diphosphate, and protein synthesis in heat-shocked *Synechococcus* sp. strain PCC 6301, a cyanobacterium. J Bacteriol 169:632-639

Swanson RV, de Lorimier R, Glazer AN (1992) Genes encoding the phycobilisome rod substructure are clustered on the *Anabaena* chromosome: Characterization of the phycoerythrocyanin operon. J Bacteriol 174:2640-2647

Takabe T, Incharoensakdi A, Arakawa K, Yokota S (1988) CO_2 fixation rate and RuBisCO content increase in the halotolerant cyanobacterium, *Aphanothece halophytica,* grown in high salinities. Plant Physiol 88:1120-1124

Tandeau de Marsac N (1991) Chromatic adaptation by cyanobacteria. In: Bogorad L, Vasil IK (eds) Cell Culture and Somatic Cell Genetics of Plants, vol 7b. Academic Press Inc, New York, pp 417-446

Tandeau de Marsac N, Houmard J (1993) Adaptation of cyanobacteria to environmental stimuli: new steps towards molecular mechanisms. FEMS Microbiol Rev 104:119-190

Tandeau de Marsac N, Mazel D, Damerval T, Guglielmi G, Capuano V, Houmard J (1988) Photoregulation of gene expression in the filamentous cyanobacterium *Calothrix* sp. PCC 7601: Light-harvesting complexes and cell differentiation. Photosynth Res 18:99-132

Tel-Or E (1980) Response of N_2-fixing cyanobacteria to salt. Appl Environ Microbiol 40:689-693

Tel-Or E, Spath S, Packer L, Mehlhorn RJ (1986) Carbon-13 NMR studies of salt shock-induced carbohydrate turnover in the marine cyanobacterium *Agmenellum quadruplicatum*. Plant Physiol 82:646-652

Terekhova IV, Chernyadyev II, Mikhailov AA, Doman NG (1986) The effect of sodium chloride on the biomass growth and on the activity of carboxylation enzymes in *Spirulina platensis*. Microbiol (Russian) 55:223-226

Thiemann B, Imhoff JF (1991) The effect of salt on the lipid composition of *Ectothiorhodospira*. Arch Microbiol 156:376-384

Thomas J, Apte SK (1984) Sodium requirement and metabolism in nitrogen-fixing cyanobacteria. J Biosci 6:771-794

Thomas SP, Shanmugasundaram S (1991) Osmoregulatory role of alanine during salt stress in the nitrogen fixing cyanobacterium *Anabaena* sp. 287. Biochem Intern 23:93-102

Tsinoremas NF, Castets AM, Harrison MA, Allen JF, de Marsac NT (1991) Photosynthetic electron transport controls nitrogen assimilation in cyanobacteria by means of posttranslational modification of the glnB gene product. Proc Natl Acad Sci 88:4565-4569

Utkilen H (1992) Cyanobacterial toxins. In: Mann NH, Carr NG (eds) Photosynthetic prokaryotes, Biotechnology handbook , vol 6. Plenum Press, New York, pp 211-231

Vega-Palas MA, Flores E, Herrero A (1992) NtcA, a global nitrogen regulator from the cyanobacterium *Synechococcus* that belongs to the Crp family of bacterial regulators. Molec Microbiol 6:1853-1859

Vierling E, Alberte RS (1980) Functional organization and plasticity of the photosynthetic unit of the cyanobacterium *Anacystis nidulans*. Physiol Plant 50:93-98

Vigh L, Gombos Z, Joo F (1985) Selective modification of cytoplasmic membrane fluidity by catalytic hydrogenation provides evidence on its primary role in chilling susceptibility of the blue-green alga, *Anacystis nidulans*. FEBS Lett 191:200-204

Völker U, Mach H, Schmid R, Hecker M (1992) Stress proteins and cross-protection by heat shock and salt stress in *Bacillus subtilis*. J Gen Microbiol 138:2125-2135

Vonshak A, Guy R, Guy M (1988a) The response of the filamentous cynobacterium *Spirulina platensis* to salt stress. Arch Microbiol 150:417-420

Vonshak A, Guy R, Poplawsky R, Ohad I (1988b) Photoinhibition and its recovery in two strains of the cyanobacterium *Spirulina platensis*. Plant Cell Physiol 29:721-726

Wada H, Gombos Z, Murata N (1990) Enhancement of chilling tolerance of a cyanobacterium by genetic manipulation of fatty acid desaturation. Nature 347:200-203

Wada H, Gombos Z, Sakamoto T, Murata N (1992) Genetic manipulation of the extent of desaturation of fatty acids in membrane lipids in the cyanobacterium *Synechocystis* PCC 6803. Plant Cell Physiol 33:535-540

Wada H, Murata N (1989) *Synechocystis* PCC 6803 mutants defective in desaturation of fatty acids. Plant Cell Physiol 30:971-978

Warr SRC, Reed RH, Chudek JA, Foster R, Stewart WDP (1985a) Osmotic adjustment in *Spirulina platensis*. Planta 163:424-429

Warr SRC, Reed RH, Stewart WDP (1984) Osmotic adjustment of cyanobacteria: the effects of NaCl, KCl, sucrose and glycine betaine on glutamine synthetase in a marine and a halotolerant strain. J Gen Microbiol 130:2169-2175

Warr SRC, Reed RH, Stewart WDP (1985b) Carbohydrate accumulation in osmotically streessed cyanobacteria (blue-green algae): Interactions of temperature and salinity. New Phytol 100:285-292

Warr SRC, Reed RH, Stewart WDP (1988) The compatibility of osmotica in cyanobacteria. Plant Cell Environ 11:137-142

Webb R, Reddy KJ, Sherman LA (1990) Regulation and sequence of the *Synechococcus* sp. strain PCC 7942 groESL operon, encoding a cyanobacterial chaperonin. J Bacteriol 172:5079-5088

Willey JM, Waterbury JB, Greenberg EP (1987) Sodium-coupled motility in a swimming cyanobacterium. J Bacteriol 169:3429-3434

Yamanaka G, Glazer AN (1981) Dynamic aspects of phycobilisome structure: Modulation of phycocyanin content of *Synechococcus* phycobilisomes. Arch Microbiol 130:23-30

Yokoyama E, Murakami A, Sakurai H, Fujita Y (1991) Effect of supra-high irradiation on the photosynthetic system of the cyanophyte *Synechocystis* PCC 6714. Plant Cell Physiol 32:827-834

Yura T, Nagai H, Mori H (1993) Regulation of the heat-shock response in bacteria. Annu Rev Microbiol 47:321-350

Zhang C-C (1993) A gene encoding a protein related to eukaryotic protein kinases from the filamentous heterocystous cyanobacterium *Anabaena* PCC 7120. Proc Natl Acad Sci 90:11840-11844

Zhao J, Brand JJ (1988) Sequential effects of sodium depletion on photosystem II in *Synechocystis*. Arch Biochem Biophys 264:657-664

Methodologies used in the Study of Nitrogen Metabolism

HELMAR ALMON and HERBERT BÖHME

Botanisches Institut der Universität Bonn, Kirschallee 1, D-5300 Bonn 1, Germany

1 Introduction

Assimilation of inorganic nitrogen by cyanobacteria involves uptake and sequential reduction of nitrate, nitrite or dinitrogen to ammonia*. Enzymes participating in these processes are nitrate

* The term "ammonia" is used throughout this article irrespective of whether the uncharged form (NH_3) or the charged form (NH_4^+) of the molecule is meant. Whenever of importance the charge of the molecule is indicated.

reductase, nitrite reductase and nitrogenase as well as carrier-proteins which mediate uptake of inorganic nitrogen compounds. Alternatively, ammonia itself can also enter cyanobacterial cells either by passive diffusion it its uncharged form (NH_3) or by active, energy dependent uptake of the NH_4^+ ion. Following reduction of oxidized N-compounds or entrance of ammonia into the cell, ammonia is immediately incorporated into C-skeletons to form amino acids. This is important because ammonia itself is highly toxic. Enzymes catalyzing assimilation of ammonia are glutamine synthetase, glutamate synthase (also called GOGAT) and glutamate dehydrogenase. Most of these enzymes have been purified from cyanobacteria; for some of them the corresponding genes have been identified and sequenced (compare also other chapters of this book).

In this chapter methods are summarized which have routinely been used to measure activities of the major enzymes of N-metabolism in cyanobacteria. In case of successful isolation of the proteins and/or the corresponding genes, methods used for these experiments will also be described. Besides a brief overview of the techniques (detailed procedures are described in the literature cited), advantages or drawbacks of single methods are critically discussed. Emphasis has been laid on more recent achievements, Nevertheless, in some cases standard procedures have been successfully used over the last years without major modifications; these will also be reviewed here.

2 Techniques for quantitative determination of N-metabolites

Some methods for quantitative determination of N-metabolites are used in connection with the study of a variety of processes discussed below. It therefore appears useful to summarize the essential techniques prior to analysis of methods dealing with individual enzyme activities. In general, it is not attempted to present detailed recipes (these can be found in the original articles) and whenever possible preference is given to methods which have been routinely and successfully used with cyanobacteria.

2.1 Determination of nitrite

The standard procedure for quantitative determination of nitrite is a colorimetric method which has been described by SNELL and SNELL (1949). A 0.5 ml sample containing up to 15 nmol NO_2^- is added to a solution consisting of 0.5 ml 1% (w/v) sulfanilamide in 2.4 M HCl and 0.5 ml 0.02% (w/v) N-(1-naphtyl)ethylenediamine hydrochloride. After an incubation period of about 15 min at room temperature, the absorbance of the diazonium dye formed is measured at 540 nm (compare also: MADUEÑO and GUERRERO 1991). This comparatively simple method allows rapid and reliable determination of nitrite and is routinely used in studies with cyanobacterial nitrite reductase and nitrate reductase as well.

2.2 Determination of nitrate

A direct determination of nitrate by UV spectroscopy has been described by CAWSE (1968) using the absorption of the nitrate ion at 210 nm. This method is simple but has some drawbacks which make it of limited use only. Any compound with high absorption in UV light contributes to a high background signal, which reduces the sensitivity of the method. Determination of nitrate uptake in the BG11 culture medium (CASTENHOLZ 1988) by this method is only possible if trace elements and ammonium iron citrate are omitted.

A colorimetric method for nitrate determination consists in the nitration of phenol-2,4-disulfonic acid or brucine (NICHOLAS and NASON 1957). In both cases an orange-brown compound is formed which can be measured by spectrophotometry. The effective range of the method is 5-100 µg of nitrate-nitrogen. The method interferes with chlorides, most forms of organic matter and high concentrations of dissolved salts.

Recently use of a nitrite reductase-less mutant strain of *Synechococcus* R2 has been shown to be an appropriate means for sensitive nitrate determination in the range of 1-50 µM nitrate (MADUEÑO and GUERRERO 1991). Whole cells of this unicellular cyanobacterium retain high levels of nitrate reductase and can be stored for months in the presence of dimethyl sulfoxide at -70°C. The nitrate photoreduction assay takes place in samples with a final volume of 1 ml containing cyanobacteria, tricine buffer (pH 8.3) and up to 50 nmol nitrate. Cells are incubated in glass tubes at 40°C with continuous shaking in the light (160 $\mu E^{-2}.s^{-1}$, white light). After reduction of nitrate to nitrite by the cyanobacteria the latter compound can be determined in the supernatant by the method of SNELL and SNELL (1949) as described above.

Use of capillary gas chromatography for nitrate determination has been described recently (DUNPHY et al. 1990); the suitability of this method for routine measurements remains still to be proven.

2.3 Determination of ammonia

There are several methods allowing rapid determination of the ammonium ion (NH_4^+). An easy to use assay is based on the Berthelot detection method, in which a blue dye complex is formed that can be quantitated spectrophotometrically. CHANEY and MARBACH (1962) and SOLORZANO (1969) have described variations of this method which is often used in cyanobacterial research. We will shortly describe an assay according to CHANEY and MARBACH (1962) which is routinely used in our laboratory. One ml of an ammonium containing solution is mixed with 0.5 ml of solution A (15 g phenol and 0.015 g Na-nitroprusside dissolved in 500 ml H_2O) and 0.5 ml solution B (10 g NaOH and 2 ml 13% NaClO dissolved in 500 ml H_2O). After incubation for 30 min in the dark the absorption of the solution is measured at 635 nm against a blank. Ammonia concentrations in the range of 10-100 µM can be measured.

Use of ammonia gas sensing electrodes has been described by TURPIN et al. (1984) and REGLINSKI et al. (1989). The principle of the method consists in addition of NaOH to the reaction medium at the end of an experiment, to convert all ammonium ions into the gaseous uncharged form, which is subsequently determined by the electrodes.

Conversion of ammonia into its *o*-phthaldialdehyde derivative and subsequent determination of the fluorescent compound by high pressure liquid chromatography (HPLC) has been found to be a fast and reliable method for ammonia determination (CORBIN 1984). This method resembles that described below for determination of glutamate and glutamine.

2.4 Determination of glutamate and glutamine

Glutamine and glutamate are substrates and/or products of the action of glutamine synthetase, glutamate synthase and glutamate dehydrogenase. Assays meassuring the activity of these enzymes often require quantitative determination of glutamine and glutamate. With HPLC having become a standard equipment in biological laboratories the use of this technique for determination of amino acids is superior to all other techniques described before. MARQUÉS et al. (1989) have improved on a method originally described by MARTIN et al. (1982) and adapted it for the study of cyanobacterial ammonia assimilating enzymes. The principle of the method consists in conversion of glutamate or glutamine into their respective *o*-phthaldialdehyde (OPA) derivatives and subsequent quantitation of the resulting compounds by reverse phase HPLC on a C_{18} column using fluorometric detection.

The experimental procedure is as follows: After stop of the enzymatic assays (for details see below), 0.4 ml of the samples are taken from the reaction mixture, centrifuged and diluted 25-fold with potassium buffer (50 mM, pH 7.5). The diluted sample (50 µl) is mixed with 150 µl of a derivatization mixture consisting of 27 mg OPA dissolved in 0.5 ml of methanol, 4.5 ml sodium borate buffer (pH 10.0), 20 µl 2-mercaptoethanol. After 90 sec incubation at 22°C, 40 µl of this assay are injected into an HPLC-apparatus. Since the kinetics of derivatization are obviously not linear and different for glutamine and glutamate, it is important to choose an exact reaction time. The actual detection of glutamate and glutamine derivatives takes place using a fluorescence detector equipped with 338 nm exitation and 425 nm emission filters.

3 Nitrate and nitrite uptake

Nitrate uptake is energy dependent and inhibited in the presence of ammonia (MARTIN-NIETO et al. 1989). A 45-47 kDa protein located within the cytoplasmic membrane seems to be involved (OMATA et al. 1989, SIVAK et al. 1989). The gene coding for this protein has been designated *nrt*A and has been sequenced recently (OMATA 1991). It is located in between the *nir* and *nar*B genes coding for nitrite reductase and nitrate reductase, respectively (LUQUE et al. 1992).

Nitrite uptake generally seems to be a passive process because it is not energy dependent (MARTIN-NIETO et al. 1989) and enters the cells in a pH-dependent manner (BÖHME 1986). Nevertheless, it has been argued that nitrite may also be transported actively by the same carrier as nitrate (RODRÍGUEZ et al. 1992).

Measurement of nitrate and nitrite uptake usually is performed by following the decrease of the respective ion concentration in the medium. Standard procedures for determination of nitrate and nitrite have been described above.

Intracellular concentrations of nitrate have been measured using acid lysates of cells which had been subjected to silicone oil centrifugation. LARA et al. (1987) describe the following procedure for *Anacystis nidulans* (*Synechococcus leopoliensis* 1402-1): An upper layer with a volume of 0.3 ml contains 25 mM Tricine/NaOH/KOH (pH 8.3),10 mM $NaHCO_3$, 50 µM KNO_3 and cells equivalent to about 10 µg chlorophyll *a*. The intermediate silicone layer consisting of a 2:1 mixture of versilube F-50 and silicone 14615-3 has a volume of 80 µl. The lower layer contains 20 µl of 2 M $H_3PO_4^-$. The reaction is started by addition of nitrate and carried out under illumination (1400 $\mu E^2.s^{-1}$) and stopped by rapid centrifugation (1.5 min, 10000 x g). Nitrate is then determined in the lower layer.

4 Ammonium transport

Ammonia is the most reduced form of N-compounds which can be used by cyanobacteria and it is preferred to all other nitrogen sources. Prior to utilization, ammonia has to be taken up by the cells. Depending on the charge of the ammonia molecule, which is dependent on the pH of the medium, uptake is either a passive or active process; e.g., in alkalophilic species, such as *Spirulina platensis* growing at pH >10, NH_3 enters cells uncharged and is trapped inside by protonation (BELKIN and BOUSSIBA 1991). Another factor influencing the rate of ammonia uptake is glutamine synthetase activity. An excellent review summarizing recent knowledge on ammonia translocation in cyanobacteria has been published by BOUSSIBA and GIBSON (1991).

Here we try to compile some important facts associated with the measurement of ammonia uptake. Prior to planning and evaluation of experiments dealing with ammonia uptake in cyanobacteria the following physicochemical parameters concerning the ammonia molecule and its often used analogue methylammonium have to be considered. The p*K*-value of ammonia is 9.25 (at 25°C); i.e., at pH 7.25 about 99% of the molecules are charged (NH_4^+), at pH 10.25 about 90% are uncharged (NH_3). The p*K*-value of methylammonium, however, is 10.65. On the other hand the permeability of methylammonium in its uncharged form is about 10 times higher than the permeability of NH_3 (P for NH_3 = 6-8 µm.s^{-1}; for CH_3NH_2 about 60-100 µm.s^{-1} [RITCHIE and GIBSON 1987]). This means that at neutral pH values the amount of uncharged ammonia or methylammonium molecules which enter the cell passively is about equal. Experiments measuring active uptake of NH_4^+ by use of methylammonium therefore do not have to consider different ratios of charged or uncharged ammonium/methylammonium molecules entering the cells.

Ammonia is very rapidly metabolized upon entrance into the cells, thus making accurate determinations of ammonia uptake difficult. It should be mentioned, however, that methylammonium, too, can be metabolized by cyanobacteria, resulting in the formation of γ-methyl glutamine (RAI et al. 1984) but no experimental evidence exists that methylammonium can be used as a nitrogen source. As pointed out by BOUSSIBA and GIBSON (1991), It is therefore essential to distinguish between free and modified methylammonium in transport studies.

Despite of all these constraints, methylammonium has been proven to be the most suitable compound for the study of ammonia uptake. For experiments the radioactively labeled [^{14}C] methylammonium is used which can be conveniently measured by following distribution of radioactivity. Assays generally are performed at neutral pH values where more than 99.9% of the added methylammonium is charged. Since the uptake process requires energy, experiments are carried out under illumination. Uptake of methylammonium in the dark is also possible, but the rates are much lower.

The use of N-isotopes in these experiments is very limited because of the short half-life of radioactive ^{13}N (about 10 min!) and because of the insensitivity of $^{15}N/^{14}N$ measurements by means of mass spectrometry, which in addition require expensive technical equipment.

5 Nitrate reductase

Cyanobacterial nitrate reductase is ferredoxin-dependent and catalyzes the reaction:

$$NO_3^- + 2Fd_{red} + H^+ \rightarrow NO_2^- + 2Fd_{ox} + H_2O$$

The enzyme has been purified from *Synechococcus* sp. (compare GUERRERO et al. 1981) and *Plectonema boryanum* (MIKAMI and IDA 1984). It consists of a single polypeptide chain with a molecular mass between 75 kDa (*Synechococcus*) and about 85 kDa (*Plectonema*). The enzyme from *Plectonema* has been shown to contain 1 atom of molybdenum as well as 4 atoms of non-heme iron and 4 acid-labile sulfur atoms. In some cyanobacteria the enzyme appears to be attached to the thylakoid membrane because prolonged sonication or use of detergents is required to solubilize nitrate reductase (HATTORI and MYERS 1967, MIKAMI and IDA 1984). Cyanobacterial nitrate reductase activity is repressed by ammonia, which has to be metabolized by glutamine synthetase, however (RAI et al. 1981, GUERRERO and LARA 1987). Induction of nitrate reductase by nitrate or nitrite has been reported for heterocystous cyanobacteria (BAGCHI et al. 1985), whereas in non-nitrogen-fixing strains nitrate and nitrite are not primarily involved in induction of enzyme levels (HERRERO et al. 1985). In *Synechococcus* sp. PCC 7942 three genes involved in reduction of nitrate (*nar*A, *nar*B, *nar*C) have been cloned (KUHLEMEIER et al. 1984a, KUHLEMEIER et al. 1984b). The *nar*B gene, possibly encoding the structural gene for nitrate reductase, has been sequenced (ANDRIESSE et al. 1990). The deduced polypeptide consists of 715 or 729 amino acids (2 possible start codons have been described) and its estimated molecular weight is about 80 kDa. The gene is located downstream from the *ntr*A gene encoding a nitrate transport polypeptide and the *nir* gene coding for nitrite reductase (LUQUE et al. 1992).

The isolation of nitrate reductase from *Synechococcus* involved cell breakage by sonication, centrifugation and as the major purification step affinity chromatography by ferredoxin-sepharose.

A standard method for *in situ* determination of nitrate reductase activity in cyanobacteria has been described by HERRERO et al. (1981). This procedure has been successfully used throughout the last decade without major modifications. The assay involves an initial permeabilization of cyanobacteria: To 1 ml of a cyanobacterial cell suspension 20 µl toluene are added and this mixture is shaken vigorously for 2 min. An aliquot of this sample is then assayed for nitrate reductase activity. Alternatively to permeabilization with toluene, mixed alkyltrimethylammonium bromide (final concentration = 50 µg/ml) has also been used for permeabilization of cyanobacteria (MARTÍN-NIETO et al. 1989). The actual reaction mixture for determination of nitrate reductase activity with a final volume of 1 ml contains $NaHCO_3/Na_2CO_3$ (100 µmol, pH 10.5), KNO_3 (20 µmol), methylviologen (4 µmol), $Na_2\ S_2O_4$ (10 µmol dissolved in 0.1 ml 0.3 M $NaHCO_3$). After a 5 min incubation at 30°C nitrite formed is determined as described above by the method of SNELL and SNELL (1949).

6 Nitrite reductase

Cyanobacterial nitrite reductase (ammonia:ferredoxin oxidoreductase, EC 1.7.7.1) is a ferredoxin-dependent enzyme catalyzing the reaction:

$$NO_2^- + 6Fd_{red} + 8H^+ \rightarrow NH_4^+ + 6Fd_{ox} + 2H_2O$$

The enzyme has been partially purified from *Anabaena cylindrica* (HATTORI and UESUGI 1968), *Anacystis nidulans* (MANZANO 1977) and *Anabaena* PCC 7119 (MÉNDEZ and VEGA 1981, MÉNDEZ et al. 1981). Enzyme preparations with high purity have been reported from *Spirulina platensis* (YABUKI et al. 1985) and *Phormidium laminosum* (ARIZMENDI and SERRA 1990). In *Anacystis* the enzyme is tightly attached to the thylakoid membrane, whereas the enzymes from *Anabaena* and *Phormidium* are soluble.

Cyanobacterial nitrite reductase has a molecular mass of about 54 kDa (a molecular mass of 68 kDa has been reported for *Anabaena cylindrica,* however) and consists of a single polypeptide chain. Similar to nitrate reductase, nitrite reductase is an inducible enzyme in heterocystous cyanobacteria. Induction of nitrite reductase in *Anabaena* is dependent on NO_2^-; NO_3^- cannot substitute for NO_2^- (OHMORI and HATTORI 1970).

The isolation of nitrite reductase from *Phormidium* as described by ARIZMENDI and SERRA (1990) involved cell disruption by sonication, anion exchange chromatography and acetone precipitation step. Further purification was achieved by FPLC using an anion exchange column (Mono-Q), hydrophobic interaction chromatography with phenyl-sepharose and gel filtration by Superose 12 columns. A 5300-fold purification was reported with an overall yield of 16% with respect to the crude extract.

Measurement of nitrite reductase activity *in vitro* can be routinely performed as described by ARIZMENDI and SERRA (1990). The reaction mixture contains in a final volume of 1 ml Tris-HCl (90 mM, pH 7.5), KNO_2 (2.5 mM), methylviologen (3 mM), sodium dithionite (20 mM, freshly dissolved in 0.3 M $NaHCO_3$) and an appropriate amount of enzyme. After 5 min preincubation the reaction is carried out at 30°C for 10 min and stopped by vigorous shaking to oxidize excess reductant. Subsequently nitrite is determined by the method of SNELL and SNELL (1949) as described above.

7 Nitrogenase

All heterocystous and several nonheterocystous (filamentous and unicellular) cyanobacteria are able to use molecular nitrogen as a nitrogen source. Dinitrogen is reduced to ammonia by the nitrogenase complex (EC 1.18.2.1), a reaction which is dependent on reduced ferredoxin and obligatorily coupled to reduction of protons resulting in formation of molecular hydrogen:

$$N_2 + 8Fd_{red} + 8H^+ + 16ATP \rightarrow NH_3 + H_2 + 8Fd_{ox} + 16ADP + 16P_i$$

Cyanobacterial nitrogenase is very similar to nitrogenases from other prokaryotes, consisting of dinitrogenase reductase (the homodimeric Fe-protein) and dinitrogenase (a heterotetrameric Mo-Fe protein). Electron flow coupled to ATP hydrolysis proceeds from ferredoxin to the Fe- protein and finally to the Mo-Fe protein. The genes coding for the structural subunits of cyanobacterial nitrogenase (*nif*HDK) have been cloned as sequenced (MEVARECH et al. 1980, MAZUR and CHUI 1982), LAMMERS and HASELKORN 1983). Nitrogenase activity and synthesis is regulated at the level of both enzyme activity and gene expression. Enzyme activity might be regulated by a post-translational modification of the Fe-protein (REICH et al. 1986, SMITH et al. 1987, REICH and BÖGER 1989).

Nitrogenase is extremely oxygen-sensitive, thus requiring special conditions for enzymatic activity (FAY 1992). In heterocystous cyanobacteria the enzyme is located exclusively with the heterocysts (WOLK and WOJCIUCH 1971). Differentiation of vegetative cells into heterocysts occurs in response to lack of utilizable nitrogen and involves sequential structural and physiological modifications necessary to provide an anaerobic environment for efficient nitrogen fixation (for review compare: HASELKORN 1978). In most heterocystous species investigated so far, a genome rearrangement during the late stage of differentiation restores *nif*HDK by excision of an interrupting 11 kb element (GOLDEN et al. 1985). The nitrogenase structural genes are subsequently transcribed from the *nif*H promoter as polycystronic message. In nonheterocystous species a temporal separation between oxygen-evolving photosynthesis and oxygen-sensitive nitrogenase activity is realized (STAL and KRUMBEIN I985, MITSUI et al. 1986).

A special ferredoxin (FdxH) is synthesized as specific electron donor to the nitrogenase complex in all N_2-fixing cyanobacteria so far tested (unpublished results of our laboratory). The occurrence and function of this ferredoxin was first described in heterocysts of *Anabaena* (SCHRAUTEMEIER and BÖHME 1985); meanwhile the corresponding gene (*fdx*H) has been identified and sequenced (BÖHME and HASELKORN 1988). The [2Fe-2S] ferredoxin has been expressed in *E. coli* (BÖHME and HASELKORN 1989) and was crystallized in order to determine Its X-ray structure (JACOBSON et al. 1992). Two lysine residues were found to be essential for interaction with nitrogenase (SCHMITZ et al. 1993).

Cyanobacterial nitrogenase has been purified from *Anabaena* strains (HALLENBECK 1979, REICH and BÖGER 1989). Preparation of the enzyme from *Anabaena variabilis* (ATCC 29413) as described by REICH and BÖGER (1989) was performed under strictly anaerobic conditions. After cell breakage with a French pressure cell and subsequent ultracentrifugation the resulting supernatant was subjected to anion exchange chromatography. The MoFe-protein and the Fe-protein were separated on DEAE-cellulose columns using Tris/HCl buffers (pH 7.8) with increasing concentrations of $MgCl_2$ for elution of the enzyme components. For elution of the Fe-protein 80 mM $MgCl_2$ was required (in contrast to 45 mM $MgCl_2$ for elution of the MoFe-protein). Further purification of the Fe-protein involved preparative SDS-PAGE.

Routine assays for determination of nitrogenase activity make use of the fact that besides N_2 several compounds with a triple bond are reduced by nitrogenase. So, acetylene (C_2H_2) is reduced to ethylene (C_2H_4). Both compounds have the advantage of being detected very easily by gas chromatography. The technical requirements necessary for an acetylene reduction assay have been described in detail previously (ALMON and BÖGER 1988). Any gas chromatograph equipped with a Porapak column and a flame ionizing detector is suitable.

For determination of *in vivo* nitrogenase activity in heterocystous cyanobacteria N_2-fixing cells (cyanobacteria grown in the absence of NO_3^- or NH_4^+) are suspended in the culture medium in reaction vessels which can be sealed with a gas-tight rubber stopper. Acetylene is added to the gas phase with a final concentration of about 10% (v/v) and in appropriate intervals aliquots from the gas phase are withdrawn and analyzed for ethylene formation. Reactions can be performed in the light or in the dark; nitrogenase activity in the dark depends on the amount of carbohydrate storage compounds present in the cells at the start of the experiment. The presence of molecular nitrogen in the reaction vessel does not interfere significantly with the experiments, because acetylene is about 60 times better soluble in aqueous solution than dinitrogen.

Induction of nitrogenase activity in nonheterocystous cyanobacteria is achieved by growth of cyanobacteria with a limiting amount of nitrate or ammonium in the culture medium (1-2 mM). After depletion of the N-source cells degrade phycobiliproteins and their colour gradually turns yellowish green. When these cells are transferred to appropriate assay conditions, nitrogenase is induced by *de novo* protein synthesis within 1-3 hours. Assays are performed in an anaerobic gas phase consisting of argon + 10%, (v/v) acetylene and under low light conditions (about 20 $\mu E.m^{-2} \cdot s^{-1}$). An equilibrium between photosynthetic oxygen evolution and respiratory oxygen uptake is achieved under these conditions and nitrogenase activity depends on respiratory activity. If higher light intensities are used, addition of DCMU (final concentration 5.10^{-6} M up to 10^{-5} M) is advantageous to prevent excess photosynthetic oxygen evolution. Alternatively, assays may also be performed in the dark with a limiting amount of oxygen present.

Nitrogenase activity in isolated heterocysts or cell-free preparations from cyanobacteria can also be determined by the acetylene reduction technique. In both cases, however, a suitable electron donor has to he used. Molecular hydrogen has been proven to be the best electron donor in heterocyst preparations if the reaction is performed in the light. In cell-free preparations an ATP-regenerating system has to be used in addition to account for the ATP requirement of the nitrogenase reaction. Sodium dithionite can be used in these systems as direct electron donor for nitrogenase. For details of electron donation to nitrogenase in heterocysts and cell-free extracts from cyanobacteria compare SCHRAUTEMEIER and BÖHME 1985, BÖHME and SCHRAUTEMEIER 1987.

Alternatively to the reduction of acetylene, the formation of molecular hydrogen can also be monitored to study nitrogenase activity. For these experiments it is necessary to exclude all possible substrates of nitrogenase from the assay. In the absence of suitable substrates nitrogenase exclusively reduces protons resulting in the evolution of H_2. For determination of hydrogen formation a gas chromatograph equipped with a thermal conductivity detector and a molecular sieve column is required (ALMON and BÖGER 1988). Due to counteracting hydrogenase activity consuming molecular hydrogen, rates of nitrogenase activity in intact cyanobacteria measured as formation of H_2 have to be analyzed in detail considering all parameters involved.

Other methods for determination of nitrogenase activity make use of N-isotopes which can be detected either by their radioactivity (^{13}N) or by mass spectrometry (^{15}N). Use of the radioactive ^{13}N-isotope has been reported for the study of cyanobacterial nitrogenase activity (WOLK et al. 1974, WOLK et al. 1976), however the extremely short half-life of the isotope (10 min) makes these experiments very inconvenient. The requirements for ^{15}N-determination have been described in detail elsewhere (JENSEN and COX 1988); if a mass spectrometer is available, this technique is promising because it allows continuous monitoring of N_2, O_2, and H_2 (JENSEN and COX 1983).

In addition to conventional molybdenum containing nitrogenase, alternative nitrogenases have been described within the last decade (for review see: PAU 1989, BISHOP and JOERGER 1990). These nitrogenases are only expressed under molybdenum deficiency and they are designated vanadium-nitrogenase or Fe-nitrogenase, respectively. The proteins are encoded by different genes called *vnf* and *anf*. In addition to the structural genes HDK a fourth component *vnf*G or *anf*G is also part of the structural genes of alternative nitrogenases. In cyanobacteria there are only few reports on the occurrence of alternative nitrogenases (KENTEMICH et al. 1988, KENTEMICH et al. 1991). A preliminary report on the identification of genes coding for an alternative nitrogenase in *Anabaena* has been presented recently (THIEL 1991).

Measurement of alternative nitrogenase activity in *Anabaena* has been described by KENTEMICH et al. (1988). Gas chromatographic analysis makes use of the fact that both alternative nitrogenases evolve small amounts of C_2H_6 in acetylene reduction assays. C_2H_6 can be detected in analogy to C_2H_4 (see above). Since alternative nitrogenases evolve more hydrogen than conventional nitrogenase does, increased rates of hydrogen evolution may also be measured by gas chromatography.

8 Glutamine synthetase

Cyanobacterial glutamine synthetase (L-glutamate:ammonia ligase, ADP-forming, EC 6.3.1.2) has a molecular weight close to 600 kDa and consists of 12 identical subunits of about 50 kDa. The reaction catalyzed is the incorporation of ammonia into glutamate and requires ATP:

$$NH_4^+ + \text{glutamate} + ATP \rightarrow \text{glutamine} + ADP + P_i$$

Glutamine synthetases have been isolated from several *Anabaena* species (STACEY et al. 1977, TULI et al. 1979, McMASTER et al. 1980, Orr et al. 1981), a *Nostoc* sp. (SAMPAIO et al. 1979), *Anacystis nidulans* (FLORENCIO and RAMOS 1985) *Phormidium laminosum* (BLANCO et al. 1989), *Calothrix* PCC 7601 (MÉRIDA et al. 1990) and *Synechocystis* PCC 6803 (MÉRIDA et al. 1990). Regulation of cyanobacterial glutamine synthetase does not involve covalent modification of the enzyme by an adenylylation/deadenylylation system as described for glutamine synthetase from enteric bacteria. Inactivation of glutamine synthetase in *Synechocystis* sp. PCC 6803 occurs via non-covalent binding of a phosphorylated compound to the enzyme (MÉRIDA 1991). The gene coding for glutamine synthetase (*Gln*A) has been identified and sequenced in *Anabaena* sp. PCC 7120 and *Calothrix* sp. PCC 7601 (TUMER et al. 1983, ELMORJANI et al. 1989). Interestingly, the gene is transcribed from different promoters depending on the available source of nitrogen. Growing with ammonia the promoter sequence resembles the consensus sequence of *E. coli* promoters, whereas under nitrogen-fixing conditions the promoter sequence is different.

Purification of glutamine synthetase from *Phormidium* as described by BLANCO et al. (1989) was performed at 4°C and involved sonic disruption of cells, centrifugation and an ammonium sulfate precipitation step of the resulting supernatant. The 45-75% saturation fraction was dialysed and subjected to anion exchange chromatography and gel filtration by Sephacryl S-300. In a final purification step the enzyme was eluted from a DE-52 column with a buffer containing a linear gradient of KCl ranging from 0.2 M to 0.4 M.

There are two different assays for determination of glutamine synthetase activity *in vitro*. Both assays have been described in detail by SHAPIRO and STADTMAN (1970) for the enzyme from *E. coli* and are also used for measurement of cyanobacterial glutamine synthetase activity:

8.1 Biosynthetic assay

$$\text{glutamate} + \text{ATP} + \text{NH}_3 \xrightarrow{\text{Mg}^{2+}/\text{Mn}^{2+}} \text{glutamine} + \text{ADP} + \text{P}_i$$

A slightly modified assay for determination of enzyme activity in *Phormidium laminosum* has been described by BLANCO et al. (1989). The reaction mixture contains in a final volume of 0.2 ml 10 µmol Tris-HCl (pH 7.2), 10 µmol L-glutamate, 10 µmol NH_4Cl, 13 µmol $MgCl_2$, 1.5 µmol ATP and an appropriate amount of the enzyme. The reaction is started by addition of ATP and the phosphate formed is determined. Alternatively, glutamate and/or glutamine can be measured by HPLC as described by MARQUÉS et al. (1989).

Biosynthetic activity of glutamine synthetase can also be determined in a coupled assay system with pyruvate kinase and lactate dehydrogenase (SHAPIRO and STADTMAN 1970).

$$\text{Glutamate} + \text{NADH} + \text{NH}_4^+ + \text{PEP} \xrightarrow{\text{ATP, Mg}^{2+}\text{, K}^+} \text{glutamine} + \text{NAD} + \text{P}_i + \text{lactate}$$

The reaction is monitored spectrophotometrically following the decrease in absorption at 340 nm.

8.2 Transferase assay

$$\text{Glutamine} + \text{NH}_2\text{OH} \xrightarrow[\text{Mg}^{2+}/\text{Mn}^{2+}]{\text{ADP, Pi, arsenate}} \gamma\text{-glutamyl hydroxamate} + \text{NH}_3$$

This is the standard assay for rapid determination of glutamine synthetase activity. The assay is usually about 5-10 times more sensitive than the biosynthetic assay and less sensitive to interference by other enzymatic activities present in cell free preparations. However, the reaction measured is of no physiological relevance. Again an assay which has been used for determination of enzyme activity in *Phormidium* (BLANCO et al. 1989) is described shortly. The reaction mix contained in a total volume of 2 ml 80 µmol Tris-HCl (pH 7.2), 60 µmol L-glutamine, 6 µmol $MnCl_2$, 0.8 µmol ADP, 120 µmol hydroxylamine (neutralized with NaOH), 37.5 µmol Na_2HAsO_4 and an appropriate amount of enzyme. The reaction is started by addition of arsenate. After addition of a stop mixture containing $FeCl_3$, trichloroacetic, and HCl the absorption at 540 nm is determined spectrophotometrically.

9 Glutamate synthase (GOGAT)

Following the incorporation of ammonia into glutamate (catalyzed by glutamine synthetase), ferredoxin-dependent glutamate synthase (EC 1.4.1.7) catalyzes the second step in ammonium assimilation:

$$\text{glutamine} + \text{2-oxoglutarate} + Fd_{red} \longrightarrow 2\ \text{glutamate} + Fd_{ox}$$

Although being a key enzyme in cyanobacterial N-metabolism, purification of glutamate synthase from a cyanobacterium has been reported only recently (MARQUÉS et al. 1992). The enzyme has been purified 2800 fold from *Synechococcus* sp. PCC 6301 with a yield of 43%. Its molecular mass is 156 kDa and it seems to be a monomer. One molecule FMN per molecule native protein has been detected. Reported K_m-values are 0.7 mM (glutamine), 1.7 mM (2-oxoglutarate), and 7 µM (ferredoxin). A cyanobacterial gene with homology to the large subunit of glutamate synthase in *E. coli* (*glt*B) has been described by LIGHTFOOT et al. (1989); however its nucleotide sequence has not yet been determined.

Isolation of glutamate synthase from *Synechococcus*, as described by MARQUÉS et al. (1992), involved: Sonic disruption of frozen cells, centrifugation and ethanol precipitation step (40-70%) of the resulting supernatant. Cation exchange chromatography and affinity chromatography with ferredoxin-sepharose were used for final enzyme purification. Since glutamate synthase was shown to be very sensitive to proteolytic degradation, the serin protease inhibitor phenylmethylsulfonyl fluoride (1 mM) had to be added. A further instability of the enzyme in solution could be prevented by addition of glutamine and 2-oxoglutarate to all buffers used.

Determination of glutamate synthase activity is possible in cell-free extracts from cyanobacteria. Enzyme activity is measured as formation of glutamate in the presence of glutamine and 2-oxoglutarate. To inhibit aminotransferase activity the inhibitor aminooxyacetate has to be added. A typical enzyme assay as described by MARQUÉS et al. (1992) for *Synechococcus* contains the following components in a final volume of 900 µl: Potassium phosphate (45 µmol, pH 7.0), L-glutamine (5 µmol), 2-oxoglutarate (1 µmol), aminooxyacetate (5 µmol), ferredoxin (10 µmol), enzyme (e.g., crude extract). The reaction is started by reduction of ferredoxin accomplished by addition of 0.8 mg sodium dithionite (dissolved in 0.1 ml of 0.12 M $NaHCO_3$). After 15 min the reaction is stopped by addition of 0.6 ml 1M HCl. Glutamate formed and residual glutamine in the reaction mixture can be measured by HPLC as described above (MARQUÉS et al. 1989).

Alternatively it is possible to use [^{14}C]glutamine as substrate and to determine [^{14}C]glutamate formed after separation of the amino acids by thin layer chromatography (RAI et al. 1982).

10 Glutamate dehydrogenase

The major pathway of cyanobacterial ammonia assimilation is undoubtedly catalyzed by the glutamine synthetase/glutamate synthase system. Nevertheless, NAD(P)H-dependent glutamate dehydrogenases (EC 1.4.1.2-4) have also been described in cyanobacteria. The reaction catalyzed is:

$$\text{2–oxoglutarate} + NH_4^+ + \text{NAD(P)H} \leftrightarrow \text{L–glutamate} + H_2O + \text{NAD(P)}^+$$

NADPH-specific glutamate dehydrogenases have been isolated from *Synechocystis* sp. PCC 6803 (FLORENCIO et al. 1987) and *Phormidium laminosum* (MARTINEZ-BILBAO et al. 1988). The enzyme from *synechocystis* has a molecular mass of 208 kDa and is obviously composed of 4 identical subunits; for *Phormidium laminosum* a molecular mass of 280 kDa has been reported; due to a contaminant protein present in this preparation no statement with respect to subunit composition could be made, however.

An NAD-specific enzyme has been described in *Synechocystis* sp. PCC 6803 (CHÁVEZ and CANDAU 1991). The molecular mass of this enzyme is 295 kDa and it seems to be a hexameric complex.

Procedures used for the isolation of NADPH-specific glutamate dehydrogenases (FLORENCIO et al. 1987, MARTINEZ-BILBAO et al. 1988) involve sonic disruption of cyanobacteria followed by centrifugation steps and anion exchange chromatography. Further purification could be achieved by chromatography on a 2,5-ADP Sepharose column and in the case of *Phormidium* by additional chromatography on a Mono-Q HR5/5 column.

Isolation of the NAD-specific enzyme (CHÁVEZ and CANDAU 1991) involved sonic disruption of cells, ammonium sulfate precipitation, chromatography on DEAE-cellulose and red-agarose columns and a final gel filtration on a Sephacryl S-300 column.

Since the reaction catalyzed by glutamate dehydrogenase is reversible, it is possible to determine either the aminating activity or the deaminating activity. A typical assay (compare FLORENCIO et al. 1987) for *in vitro* measurement of the aminating activity of NADPH-specific glutamate dehydrogenase contains 85 mM Tris-HCl (pH 8.0), 10 mM 2-oxoglutarate, 50 mM NH_4Cl and 0.2 mM NADPH. The reaction is started by addition of ammonium chloride and can be followed spectrophotometrically measuring the decrease in absorbance at 340 nm (including controls for ammonia-independent oxidation of NADPH) or by determination of glutamate formed. For *in situ* assays with *Synechocystis* 6803 cells may be permeabilized with mixed alkyltrimethylammonium bromide (final concentration = 0.025%). Deaminating activity can be followed spectrophotometrically at 340 nm in a reaction mixture consisting of 80 mM Tris-HCl (pH 9.0), 50 mM L-glutamate and 2 mM $NADP^+$.

The activity of the NAD-specific enzyme from *Synechocystis* 6803 has been determined by similar assay systems (compare CHÁVEZ and CANDAU 1991); NADP(H) was of course substituted by NAD(H).

References

Almon H, Böger P (1988) Nitrogen and hydrogen metabolism: Induction and measurement. Meth Enzymol 167:459-467

Andriesse X, Bakker H, Weisbeek P (1990) Analysis of nitrate reduction genes in cyanobacteria. In: Ullrich WR, Rigano C, Fuggi A, Aparicio PJ (eds) Inorganic nitrogen in plants and microorganisms, Springer Verlag, Berlin, pp 303-307

Arizmendi JM, Serra JL (1990) Purification and some properties of the nitrite reductase from the cyanobacterium *Phormidium laminosum.* Biochim Biophys Acta 1040:237-244

Bagchi SN, Rai AN, Singh HN (1985) Regulation of nitrate reductase in cyanobacteria. Repression-derepression control of nitrate reductase apoprotein in the cyanobacterium *Nostoc muscorum.* Biochim Biophys Acta 838:370-373

Belkin S, Boussiba S (1991) Resistance of *Spirulina platensis* to ammonia at high pH values. Plant Cell Physiol 32:953-958

Bishop PE, Joerger RD (1990) Genetics and molecular biology of alternative nitrogen fixation systems. Annu Rev Plant Physiol Plant Mol Bio 41:109-125

Blanco F, Alaña A, Llama MJ, Serra JL (1989) Purification and properties of glutamine synthetase from the non-N_2-fixing cyanobacterium *Phormidium laminosum.* J Bacteriol 171:1158-1165

Böhme H (1986) Inhibition of nitrogen fixation by nitrite in *Anabaena variabilis.* Arch Microbiol 146:99-103

Böhme H, Haselkorn R (1988) Molecular cloning and nucleotide sequence analysis of the gene coding for heterocyst ferredoxin from the cyanobacterium *Anabaena* sp. strain PCC 7120. Mol Gen Genet 214:278-285

Böhme H, Haselkorn R (1989) Expression of *Anabaena* ferredoxin genes in *Escherichia coli.* Plant Mol Biol 12:667-672

Böhme H, Schrautemeier B (1987) Electron donation to nitrogenase in a cell-free system from heterocysts of *Anabaena variabilis.* Biochim Biophys Acta 891:115-120

Boussiba S and Gibson J (1991) Ammonia translocation in cyanobacteria. FEMS Microbiol Rev 88:1-14

Castenholz RW (1988) Culturing methods for cyanobacteria. Meth Enzymol 167:68-93

Cawse PA (1968) The determination of nitrate in soil solutions by ultraviolet spectrophotometry. The Analyst 92:311-315

Chaney AL, Marbach EP (1962) Modified reagents for the determination of urea and ammonia. Clin Chem 8:130-132

Chávez S, Candau P (1991) An NADH specific glutamate dehydrogenase from cyanobacteria. Identification and properties. FEBS Lett 285:35-38

Corbin JL (1984) Liquid chromatographic-fluorescence determination of ammonia from nitrogenase reactions: A 2-min assay. Appl Environm Microbiol 47:1027-1030

Dunphy MJ, Goble DD, Smith DJ (1990) Nitrate analysis by capillary gas chromatography. Anal Biochem 184:381-387

Elmorjani K, Liotenberg S, Houmard J, Tandeau de Marsac N (1989) Molecular characterization of the gene encoding glutamine synthetase in the cyanobacterium *Calothrix* sp. PCC 7601. Biochem Biophys Res Commun 189:1296-1302

Fay P (1992) Oxygen relations of nitrogen fixation in cyanobacteria. FEMS Microbiol Rev 56:340-373

Florencio FJ, Marqués S, Candau P (1987) Identification and characterization of a glutamate dehydrogenase in the unicellular cyanobacterium *Synechocystis* PCC 6803. FEBS Lett 223:37-41

Florencio FJ, Ramos JL (1985) Purification and characterization of glutamine synthetase from the unicellular cyanobacterium *Anacystis nidulans*. Biochim Biophys Acta 838:39-48

Golden JW, Robinson SJ, Haselkorn R (1985) Rearrangement of nitrogen fixation genes during heterocyst differentiation in the cyanobacterium *Anabaena*. Nature 314:419-423

Guerrero MG, Lara C (1987) Assimilation of nitrogen. In: Fay P, van Baalen C (eds) The cyanobacteria. Elsevier Science Publishers, Amsterdam, pp 163-186

Guerrero MG, Vega JM, Losada M (1981) The assimilatory nitrate-reducing system and its regulation. Annu Rev Plant Physiol 32:169-204

Hallenbeck PC. Kostel PJ, Benemann JR (1979) Purification and properties of nitrogenase from the cyanobacterium, *Anabaena cylindrica*. Eur J Biochem 98:275-284

Haselkorn R (1978) Heterocysts. Annu Rev Plant Physiol 29:319-344

Hattori A, Myers J (1967) Reduction of nitrate and nitrite by subcellular preparations of *Anabaena cylindrica*. II. Reduction of nitrate to nitrite. Plant Cell Physiol 8:327-337

Hattori A, Uesugi I (1968) Purification and properties of nitrite reductase from the blue-green alga *Anabaena cylindrica*. Plant Cell Physiol 9:689-699

Herrero A, Flores E, Guerrero MG (1981) Regulation of nitrate reductase levels in the cyanobacteria *Anacystis nidulans, Anabaena* sp. strain 7119, and *Nostoc* sp. strain 6719. J Bacteriol 145:175-180

Herrero A, Flores E, Guerrero MG (1985) Regulation of nitrate reductase cellular levels in the cyanobacteria *Anabaena variabilis* and *Synechocystis* sp. FEMS Microbiol Lett 26:21-25

Jacobson BL, Chae YK, Böhme H, Markley JL, Holden HM (1992) Crystallization and preliminary analysis of oxidized, recombinant, heterocyst [2Fe-2S] ferredoxin from *Anabaena* 7120. Arch Biochem Biophys 294:279-281

Jensen BB, Cox RP (1983) Direct measurements of steady-state kinetics of cyanobacterial N_2 uptake by membrane-leak mass spectrometry and comparisons between nitrogen fixation and acetylene reduction. Appl Environm Microbiol 45:1331-1337

Jensen BB, Cox RP (1988) Measurement of hydrogen exchange and nitrogen uptake by mass spectrometry. Meth Enzymol 167:467-474

Kentemich T, Danneberg G, Hundeshagen B, Bothe H (1988) Evidence for the occurrence of the alternative, vanadium-containing nitrogenase in the cyanobacterium *Anabaena variabilis*. FEMS Microbiol Lett 51:19-24

Kentemich T, Haverkamp G. Bothe H (1991) The expression of a third nitrogenase in the cyanobacterium *Anabaena variabilis*. Z Naturforsch 46c:217-222

Kuhlemeier CJ, Logtenberg T, Stoorvogel W, van Heugten HAA, Borrias WE, van Arkel GA (1984a) Cloning of nitrate reductase genes from the cyanobacterium *Anacystis nidulans*. J Bacteriol 159:36-41

Kuhlemeier CJ, Teeuwsen VJP, Janssen MJT, van Arkel GA (1984b) Cloning of a third nitrate reductase gene from the cyanobacterium *Anacystis nidulans* R2 using a shuttle cosmid library. Gene 31:109-116

Lammers P, Haselkorn R (1983) Sequence of the *nif*D gene coding for the α-subunit of dinitrogenase from the cyanobacterium *Anabaena*. Proc Natl Acad Sci USA 80:4723-4727

Lara C, Romero JM Guerrero MG (1987) Regulated nitrate transport in the cyanobacterium *Anacystis nidulans*. J Bacteriol 169:4376-4378

Lightfoot DA. McPherson MJ, Wootton JC (1989) Detection of a homologue to an *E. coli* glutamate synthase gene in a cyanobacterium. FEMS Microbiol Lett 57:129-132

Luque I, Herrero A, Flores E, Madueño F (1992) Clustering of genes involved in nitrate assimilation in the cyanobacterium *Synechococcus*. Mol Gen Genet 232:7-11

Madueño F, Guerrero MG (1991) Use of a mutant strain of the cyanobacterium *Synechococcus* R2 for the determination of nitrate. Anal Biochem 198:200-202

Manzano C (1977) La reducción fotosintética del nitrato en El alga verdeazulada Anacystis nidulans. PhD thesis, University of Sevilla, Sevilla

Marqués S, Florencio FJ, Candau P (1989) Ammonia assimilating enzymes from cyanobacteria: In situ and in vitro assay using high performance liquid chromatography. Anal Biochem 180:152-157

Marqués S, Florencio FJ, Candau P (1992) Purification and characterization of the ferredoxin-glutamate synthase from the unicellular cyanobacterium *Synechococcus* sp. PCC 6301. Eur J Biochem 206:69-77

Martin F, Suzuki A, Hirel B (1982) A new high-performance liquid chromatography assay for glutamine synthetase and glutamate synthase in plant tissues. Anal Biochem 125:24-29

Martinez-Bilbao M, Martinez A, Urkijo I, Llama MJ, Serra JL (1988) Induction, isolation, and some properties of the NADPH- dependent glutamate dehydrogenase from the nonheterocystous cyanobacterium *Phormidium laminosum*. J Bacteriol 170:4897-4902

Martín-Nieto J, Herrero A, Flores E (1989) Regulation of nitrate and nitrite reductases in dinitrogen-fixing cyanobacteria and Nif^- mutants. Arch Microbiol 151:475-478

Mazur BJ, Chui CF (1982) Sequence of the gene coding for the a- subunit of dinitrogenase from the blue-green alga *Anabaena*. Proc Natl Acad Sci USA 79:6782-6786

McMaster BJ, Danton MS, Storch TA, Dunham VL (1980) Regulation of glutamine synthetase in the blue-green alga *Anabaena flos-aquae*. Biochem Biophys Res Commun 96:975-983

Méndez JM, Herrero A, Vega JM (1981) Characterization and catalytic properties of nitrite reductase from *Anabaena* sp. 7119. Z Pflanzenphysiol 103:305-315

Méndez JM. Vega JM (1981) Purification and molecular properties of nitrite reductase from *Anabaena* sp. 7119. Physiol Plant 5:2:7-14

Mérida A, Candau P, Florencio FJ (1991) *In vitro* reactivation of *in vivo* ammonium-inactivated glutamine synthetase from *Synech*ocystis sp. PCC 6803. Biochem Biophys Res Commun 181:780-786

Mérida A, Leurentop L, Candau P, Florencio FJ (1990) Purification and properties of glutamine synthetases from the cyanobacteria *Synechocystis* sp. strain PCC 6803 and *Calothrix* sp. strain PCC 7601. J Bacteriol 172:4732-4735

Mevarech M, Rice D, Haselkorn R (1980) Nucleotide sequence of a cyanobacterial *nif*H gene coding for nitrogenase reductase. Proc Natl Acad Sci USA 77:6476-6480

Mikami B. Ida S (1984) Purification and properties of ferredoxin-nitrate reductase form the cyanobacterium *Plectonema boryanum*. Biochim Biophys Acta 791:294-304

Mitsui A, Kumazawa S, Takahashi A, Ikemoto H, Cao S, Arai T (1986) Strategy by which nitrogen-fixing unicellular cyanobacteria grow photoautotrophically. Nature 323:720-722

Nicholas DJ, Nason A (1957) Determination of nitrate and nitrite. Meth Enzymol 3:981-984

Ohmori M. Hattori A (1970) Induction of nitrate and nitrite reductases in *Anabaena cylindrica*. Plant cell Physiol 11: 873-878

Omata T (1991) Cloning and characterization of the *nrt*A gene that encodes a 45 kDa protein involved in nitrate transport in the cyanobacterium *Synechococcus* PCC 7942. Plant Cell Physiol 32:151-157

Omata T, Ohmori M, Arai N, Ogawa T (1989) Genetically engineered mutant of the cyanobacterium *Synechococcus* PCC 7942 defective in nitrate transport. Proc Natl Acad Sci USA 86:6612-6616

Orr J, Keefer LM, Kein P, Dinh Nguyent T, Wellems T, Heinrikson RL, Haselkorn R (1981) Purification, physical characterization and NH_2-terminal sequence of glutamine synthetase from the cyanobacterium *Anabaena* 7120. J Biol Chem 256:13091-13098

Pau RN (1989) Nitrogenases without molybdenum. TIBS 14:183-186

Rai AK, Kashyap AK, Gupta SL (1981) ATP-dependent uptake of nitrate in *Nostoc muscorum* and inhibition by ammonium ions. Biochim Biophys Acta 674:78-86

Rai AN, Rowell P, Stewart WDP (1982) Glutamate synthase activity of heterocysts and vegetative cells of the cyanobacterium *Anabaena variabilis* Ktz. J Gen Microbiol 128:2203-2205

Rai AN, Rowell P, Stewart WDP (1984) Evidence for an ammonium transport system in free-living and symbiotic cyanobacteria. Arch Microbiol 137:241-246

Reglinski A, Rowell P, Kerby NW, Stewart WDP (1989) Characterization of the methylammonium/ammonium transport system in mutant strains of *Anabaena variabilis* resistant to ammonium analogues. J Gen Microbiol 135:1441-1451

Reich S, Almon H, Böger P (1986) Short-term effect of ammonia on nitrogenase activity of *Anabaena variabilis* (ATCC 29413) FEMS Microbiol Lett 34:53-56

Reich S, Böger P (1989) Regulation of nitrogenase activity in *Anabaena variabilis* by modification of the Fe protein. FEMS Microbiol Lett 58:81-86

Ritchie RJ and Gibson J (1987) Permeability of ammonia, methylamine and ethylamine in the cyanobacterium *Synechococcus* R- 2 (*Anacystis nidulans*) PCC 7942. J Membr Biol 95:131-142

Rodríguez R, Lara C, Guerrero MG (1992) Nitrate transport in the cyanobacterium *Anacystis nidulans* R2: Kinetic and energetic aspects. Biochem J 282:639-643

Sampaio MJ, Rowell P, Stewart WDP (1979) Purification and some properties of glutamine synthetase from the nitrogen-fixing cyanobacteria *Anabaena cylindrica* and a *Nostoc* sp. J Gen Microbiol 111:181-191

Schmitz S, Schrautemeier B, Böhme H (1993) Evidence from directed mutagenesis that positively charged amino acids are necessary for interaction of nitrogenase with the [2Fe-2S] heterocyst ferredoxin (FdxH) from the cyanobacterium *Anabaena* sp. PCC 7120. Mol Gen Genet 240:455-460

Schrautemeier B, Böhme H (1985) A distinct ferredoxin for nitrogen fixation isolated from heterocysts of the cyanobacterium *Anabaena variabilis*. FEBS Lett 184:304-308

Shapiro BM, Stadtman ER (1970) Glutamine synthetase (*Escherichia coli*) Meth Enzymol 17:910-922

Sivak MN, Lara C, Romero JM, Rodríguez R, Guerrero MG (1989) Relationship between a 47 kDa cytoplasmic membrane polypeptide and nitrate transport in *Anacystis nidulans*. Biochem Biophys Res Commun 158:257-262

Smith RL, van Baalen C, Tabita FR (1987) Alteration of the Fe protein of nitrogenase by oxygen in the cyanobacterium *Anabaena* sp. strain CA. J Bacteriol 169:2537-2542

Snell FD, Snell CT (eds) (1949) Colorimetric methods of analysis, vol 2. van Nostrand, New York, pp 804-805

Solorzano L (1969) Determination of ammonia in natural sea water by the phenolhypochloride method. Limnol Oceanogr 14:799-801

Stacey G, Tabita FR, van Baalen C (1977) Nitrogen and ammonia assimilation in the cyanobacteria: purification of glutamine synthetase from *Anabaena* sp. strain CA. J Bacteriol 132:596-603

Stal LJ, Krumbein WE (1985) Nitrogenase activity in the non- heterocystous cyanobacterium *Oscillatoria* sp. grown under alternating light-dark cycles. Arch Microbiol 143:67-71

Thiel T (1991) Isolation and characterization of genes for an alternative nitrogenase in the cyanobacterium *Anabaena variabilis* ATCC 29413. VII International Symposium on Photosynthetic Prokaryotes, Abstract 152B

Tuli R, Jawali N, Thomas J (1979) A rapid method for the purification of glutamine synthetase from the blue-green alga *Anabaena* L-31. Ind J Exp Biol 17:1239-1241

Tumer NE, Robinson SJ, Haselkorn R (1983) Different promoters for the *Anabaena* glutamine synthetase gene during growth using molecular or fixed nitrogen. Nature 306:357-342

Turpin DH, Edie SA, Canvin DT (1984) *In vivo* nitrogenase regulation by ammonium and methylamine and the effect of MSX on ammonium transport in *Anabaena flos-aquae*. Plant Physiol 74 701-704

Wolk CP, Austin SM, Bortins J, Galonsky A (1974) Autoradiographic localization of ^{13}N after fixation of ^{13}N-labeled nitrogen gas by a heterocyst-forming blue-green alga. J Cell Biol 61:440-453

Wolk CP, Thomas J, Shaffer W, Austin SM, Galonsky A (1976) Pathway of nitrogen metabolism after fixation of ^{13}N-labeled nitrogen gas by the cyanobacterium *Anabaena Cylindrica*. J Biol Chem 251:5027-5034

Wolk CP, Wojciuch E (1971) Photoreduction of acetylene by heterocysts. Planta 97:126-134

Yabuki Y, Mori E, Tamura G (1985) Nitrite reductase in the cyanobacterium *Spirulina platensis*. Agric Biol Chem 49:3061-3062

Future prospects

IOAN I. ARDELEAN

Institute of Biology, Splaiul Independentei 296, Bucharest 79651, POB 17-50, Romania

1 Introduction

During the last two decades cyanobacteria have been studied extensively (CARR and WHITTON 1973, FOGG et al. 1973, STEWART and GALLON 1982, FAY and VAN BAALEN 1987, PACKER and GLAZER 1988, DEWS 1990, BRYANT 1994) especially after the proposal of STANIER et al. (1978) for their inclusion to the bacterial kingdom, which gained acceptance throughout the international scientific community.

Cyanobacteria are photosynthetic prokaryotes commonly found in different natural habitats, both terrestrial and aquatic. They are the highly successful group of organisms regarding their ecological distribution because of their biological properties. The most important being:

a) They can use water as electron donor through the operation of two distinct photosystems, namely photosystem 1 (PS1) and photosystem 2 (PS2), as in photosynthetic eukaryotes.

b) During anoxic conditions, after an induction period, cyanobacteria can reversibly carry out a PS1-dependent anoxygenic photosynthesis, using sulphide, thiosulphate or H_2, as electron donor instead of water.

c) They can maximise light absorption by changing the chemical composition of light harvesting pigments phycocyanin, allophycocyanin and phycoerythrin (complementary chromatic adaptation), having absorption maxima at 650, 620 and 565 nm, respectively.

d) They can balance the electron flow between PS1 and PS2, thereby protecting PS2 and maximising the energy flow within the photosynthetic chain (S1-S2 state transitions).

e) They can use atmospheric molecular nitrogen (N_2) as sole nitrogen source for biosynthetic purposes, thereby being able to populate environments devoid of nitrogenous compounds, organic or inorganic.

f) Their endogenous aerobic respiration proceeds at rates comparable to those of eukaryotic algae; however the increase in respiratory rates following substrate addition is significant only in those species able to perform heterotrophic or mixotrophic growth.

g) Some species can grow under dark anaerobic conditions by the operation of fermentative breakdown of carbohydrates.

h) Cyanobacteria are able to deposit different reserve compounds (glycogen, poly-β-hydroxibutyrate, polyphosphate, cyanophycin granules) inside the cell, which are important for survival during adverse conditions.

i) Some freshwater cyanobacteria develop salt tolerance (0.5 - 1.0 M NaCl) by involving a mechanism of co-operation between respiratory electron transfer and photosynthetic electron transfer for the extrusion of Na^+ outside the cell.

j) Many cyanobacteria are able to differentiate cells serving for protection, perennation, reproduction or nitrogen-fixation.

k) Some cyanobacteria are motile; moreover, the planktonic forms can adjust to their optimum position within the water column using buoyancy provided by the gas vesicles.

l) Some cyanobacteria can form symbiotic association with algae, fungi, liverworts, ferns and higher plants, while others are engaged in symbiosis with animals providing them with photosynthate and nitrogen.

m) Some species liberate toxins.

Their capability to synthesise organic chemicals from relatively simple and widespread raw materials (water, carbon dioxide, ammonium, nitrate) at the expense of light energy made them extremely useful living cells for fundamental and biotechnology-oriented research. Cyanobacteria are unique biological material to study the following main aspects in prokaryotes, related to the subject of this book:

i) The interaction between nitrogen-fixation and oxygenic photosynthesis is one of the most exciting topics, because photosynthesis evolves oxygen while nitrogen-fixation is an extremely oxygen-sensitive process; this is why cyanobacteria have developed different strategies to allow the occurrence of both processes in the same organism, either unicellular or filamentous.

ii) The interaction between nitrogen-fixation and aerobic respiration is also important because aerobic respiration decreases the level of molecular oxygen, thereby protecting nitrogenase; moreover this catabolic process is involved in providing useful energy for nitrogenase function.

iii) The interaction between nitrate assimilation and oxygenic photosynthesis.

iv) The reversible transition from oxygenic photosynthesis to anoxygenic photosynthesis; under these conditions they can also fix molecular nitrogen if no other source of nitrogen is present.

v) The interplay between aerobic respiration and oxygenic photosynthesis.

vi) Being one of a few group of prokaryotes able to differentiate specialised cells cyanobacteria are also useful to study all the above aspects during cell differentiation.

This chapter presents some of the strategies proposed to manage cyanobacterial nitrogen metabolism and related processes to improve the usefulness of cyanobacteria for environmental biotechnology. Firstly, attention will be focused on the lesson we could learn from the way by which Nature enables both heterocystous and non-heterocystous cyanobacteria to fix nitrogen (2.1-2.2). Secondly, some proposals (2.3-2.6) will be introduced that could improve their nitrogen metabolism, with special emphasis on nitrogen-fixation. Thirdly, the above proposals (2.1-2.6) will be discussed within the framework of six application (3.1-3.6) of cyanobacteria in the field of environmental biotechnology.

2 Proposals

2.1 N_2-fixation in nonheterocystous cyanobacteria

In nonheterocystous cyanobacteria nitrogen-fixation (NIF) occurs within the same cell together with photosynthesis and respiration, the other two important energetic processes. In these cyanobacteria aerobic respiration is the main source of ATP and reducing equivalents required for nitrogen-fixation, but ultimately the source of respiratory substrates is photosynthesis. From these strategies developed by cyanobacteria to allow the occurrence of both processes in the same cell we could learn more about the flexibility of prokaryotes that enables them to grow in a hostile environment. Moreover it is possible, and probably very important both for fundamental and biotechnological research, to try to improve these strategies within the same organism or to transfer them, either improved or not, to other cyanobacteria. The main strategy developed by nonheterocystous cyanobacteria to sustain NIF is based on temporal separation between NIF and oxygenic photosynthesis (GALLON et al. 1974, MULLINEAUX et al. 1981). Usually this segregation follows the diurnal cycle, with NIF reaching a maximum during the dark period and photosynthetic activity occurring in the light period (STAL and KRUMBEIN 1981, 1987). However, during the yester years few important findings have emerged.

2.1.1 *Synechococcus*

MITSUI and co-workers (LEON et al. 1986, KUMAZAWA and Mitsui 1992) in synchronised cells of *Synechococcus* strain Miami BG 43511, incubated under alternating 12 hours light dark cycles found photosynthesis peak in the light, whereas nitrogenase activity was measured in the middle of the dark period together with an increase in the respiratory activity. The very interesting aspect is the maintenance of this temporal separation during incubation of already synchronised cells under continuous light as long as 60 hours. This means that synchronised cells of *Synechococcus* strain Miami 43511 display a temporal separation of photosynthesis over nitrogenase and respiration even when the light is supplied continuously. Furthermore, the general assessment is that in cyanobacteria and other prokaryotes as (e.g., Rhodospirillaceae, photosynthesis and respiration interact with each other, the best known result of this interplay being the inhibition of respiration by light (VAN NIEL 1941, JONES and MYERS 1963, MCCARTHY and FERGUSON 1982, RUGOLO and ZANNONI 1983, SCHERER et al. 1988). It seems that this strain exhibits a mechanism that enables it to increase respiratory activity in light.

The knowledge of this mechanism will enable us to have a better understanding of photosynthesis-respiration interplay in cyanobacteria either nitrogen-fixing or not. If we could learn how to manage this interplay it is possible to improve the nitrogen-fixing ability significantly of both the heterocystous and nonheterocystous cyanobacteria. It is further interesting to control the temporal separation of these processes in continuous light in cyanobacteria.

First, in unicellular and vegetative cells of nonheterocystous cyanobacterial species with nitrogenase but unable of NIF under normal oxygen pressure, it would be possible to promote nitrogenase activity at reasonable rates. Second, species without nitrogenase could be the subject of genetic manipulation to transfer the ability to encode and synthesise this enzymatic complex and further to manage temporal separation of nitrogenase and respiration over photosynthesis. Third, vegetative cells of heterocystous cyanobacteria could become capable to synthesise nitrogenase and fix nitrogen actively.

2.1.2 *Trichodesmium*

In cells of *Trichodesmium* spp. a linear correlation between the quantity of nitrogenase and that of cytochrome oxidase synthesised by different positioned cells within the colony, clearly shows the relationship between NIF and respiratory activity (BERGMAN et al. 1993). It would be interesting to be able to increase cytochrome oxidase and corresponding respiratory activity to enhance NIF in cyanobacteria, either unicellular or filamentous, and their capability to grow in nitrogen-free environments.

2.1.3 *Lyngbya*

In *Lyngbya aestuarii*, nitrogenase is localised mainly at the two ends of the filament in those cells where less thylakoids are present. This seems to be a spatial separation between NIF and photosynthesis (PAERL et al. 1991). It would be useful to promote a spatial separation of these processes in other filamentous nonheterocystous cyanobacteria which may further improve the ability to fix N_2.

2.2 N_2-fixation in heterocystous cyanobacteria

Heterocystous cyanobacteria are uniquely gifted for NIF because their nitrogenase activity occurs in specialised cells, the heterocysts.

Thus, it seems promising to look after some of structure-function relationship in these cyanobacteria to find out suggestions to enhance their ability to use N_2 as sole nitrogen source.

2.2.1 Heterocyst envelope

Heterocyst envelope contains three layers (the fibrous, the homogenous and the laminated). These layers are deposited outside the cell wall during differentiation of a vegetative cell into a mature heterocyst (see FAY 1992). These layers, specially the homogenous and laminated ones, are important for the protection of nitrogenase against oxygen. The mutants of *Anabaena variabilis* strain ATCC 29413 deficient in the laminated layer are unable to fix nitrogen under aerobic conditions. The major components of this layer are three glycolipids that seem to act as a barrier against oxygen (MURRY and WOLK 1989). It would be interesting to induce the synthesis of these glycolipids into a vegetative cell to obtain a modified vegetative cell (MVC) concerning oxygen permeability; under these protective conditions it would be possible to promote nitrogenase synthesis and activity or to enhance an already occurring activity in vegetative cells of unicellular and filamentous cyanobacteria.

These MVC having one or more type of glycolipids are important because:

i) They could be used further to study the role of each type of glycolipids in the protection of nitrogenase, and their co-operation, if any.

ii) To design new glycolipids better suited to sustain nitrogenase activity in cyanobacteria.

iii) They could be interesting biological material to study photosynthesis, respiration and their interplay under limited oxygen diffusion inside and outside the cell, that could further improve the usefulness of cyanobacteria for environmental biotechnology.

iv) They could be a living model to study the differentiation of heterocysts whether a fully developed heterocyst can be obtained or not; moreover, these MVC should be more successful than unmodified cells regarding the promotion of nitrogenase synthesis and activity.

All these approaches with vegetative cells whether successful or not concerning practical applications, would improve our knowledge immensely regarding the structural-functional relationships in heterocysts. This knowledge could enable us to design new types of heterocysts with respect to:

a) chemical composition of the envelope;

b) optimised thickness of each layer and

c) better co-operation between reducing equivalents, ATP and nitrogen supplies on one hand and ammonia and related compounds withdrawn on the other hand, in order to improve NIF activity both in free and symbiotic nitrogen-fixing cyanobacteria.

2.2.2 Absence of PS2 in heterocysts

Heterocysts do not evolve oxygen as they are unable to carry out oxygenic photosynthesis. The PS2 activity is absent in these cells (BRADLEY and CARR 1971, DONZE et al. 1972, TEL-OR and STEWART 1975). PS1-dependent anoxygenic photosynthesis generates ATP for nitrogenase activity (STEWART et al. 1975, HOUCHINS and HIND 1983). Based on the fact that PS2 deficient mutants of cyanobacteria have already been obtained (SHELLY and VERMAAS 1989, VERMAAS et al. 1990), it is worth to obtain this kind of mutants in nonheterocystous cyanobacteria to study the followings:

i) Nitrogenase synthesis and activity,if any, in cells where the endogenous source of oxygen is abolished and/or where the permeability for external oxygen is restricted by proper synthesis of suitable layers.

ii) To study the relationship between photosynthesis, respiration and NIF in these cells.

iii) To use these cells as an experimental model for the differentiation of a vegetative cell into a mature heterocyst.

2.3 Ferredoxin

An increase in the concentration of ferredoxin, the physiological electron donor to nitrogenase, could enhance nitrogenase activity, as well as other ferredoxin-dependent processes (e.g., nitrate assimilation). Chemical modification of ferredoxin molecule in order to enhance its affinity both at reducing site and at oxidising site could improve not only NIF but also other important processes (e.g., molecular hydrogen production and nitrate assimilation). Nitrogen metabolism being intimately connected both with photosynthesis and respiration, it is expected that any change in the rate of these two bioenergetic processes as a result of an increase in the concentration of a photosynthetic or respiratory electron carrier would affect NIF as well as the nitrogen cycle within the cell.

2.4 Nitrogenase

Nitrogenase is an enzyme system composed by two proteins, dinitrogenase and dinitrogenase reductase, which lacks an absolute specificity for molecular nitrogen. This is why during function it also reduces protons, as well as other substrates (see BURRIS 1991). Biochemical and genetic engineering of nitrogenase, mainly of the protein responsible for protons reduction in order to obtain a modified enzymic complex unable to reduce protons could be useful to improve NIF efficiency; under these conditions no ATP and reducing equivalents will be consumed for hydrogen production, consequently the energetic efficiency of the process is highly increased. Conversely, a modified enzyme unable to reduce nitrogen but keeping its ability to reduce protons could be used to enhance hydrogen production in cyanobacteria as well as in other bacteria.

A modified nitrogenase would also be useful to study structure-function relationship in these proteins and to manage their catalytic sites. These engineered proteins could be introduced into other cyanobacteria and bacteria to increase NIF. It may prove to be a better biological model for biomimetic systems able to perform one or more of the reactions catalysed by the nitrogenase, but further problems may arise as follows:

i) Does the increased NIF activity fit the general metabolism?

ii) Does the cyanobacteria with increased NIF grow well in lab or in outdoor tanks or even in natural ecosystems?

iii) What about ecological implications of releasing such engineered cyanobacteria in environment?

The increased NIF activity should fit well with the pathways of ammonia assimilation, otherwise the excess of nonassimilated ammonia will be found in the medium as ammonium ion; fortunately, these cells may be used as fertilisers in agriculture or to produce aminoacids to be used as food additives for human or animals, or as raw materials for other biotechnologies.

2.5 Mutants derepressed for nitrogenase

Of utmost importance for environmental application is to obtain strains of nitrogen-fixing cyanobacteria derepressed for nitrogenase, thereby being able to fix nitrogen even when N-sources other than N_2 are present.

2.6 Hydrogenase

Hydrogenases are the enzymes responsible for the reversible reduction of protons. According to the reaction catalysed *in vivo* there are two kinds of hydrogenases:

i) an unidirectional uptake hydrogenase (Hup) and

ii) a reversible hydrogenase.

In cyanobacteria, Hup activity occurs only under diazotrophic growth (EISBRENER et al. 1978, HOUCHINS and BURRIS 1981a, b) besides the immunolocalization of protein both in heterocysts and vegetative cells (LINDBLAD and SELLSTEDT 1990). According to FAY (1992), Hup appears to be universally present in aerobic nitrogen-fixing organisms. Hup is involved in consumption of some of the H_2 evolved during nitrogen-fixation (PETERSON and BURRIS 1978). This reaction mediated by the respiratory electron transport chain occurs with the concomitant reduction of oxygen. The use of H_2 as respiratory substrate (oxy-hydrogen or Knall gas reaction) can reach the rate of endogenous respiration (BOTHE et al. 1982) arguing again the involvement of respiration in nitrogen metabolism of cyanobacteria. The oxy-hydrogen reaction increases the efficiency of NIF as follows:

i) It generates ATP (PETERSON and BURRIS 1978).

ii) It consumes oxygen lowering its partial pressure at the cellular level, thereby augmenting the effect of respiratory protection (DIXON 1972).

iii) It removes H_2 which is inhibitory for nitrogen reduction (MORTENSON 1978).

The increase in Hup activity in isolated heterocysts in light and in the presence of suitable electron acceptors (e.g., NADP, methyl viologen) (HOUCHINS and BURRIS 1981a) argues the involvement of photosynthetic electron transport in hydrogen consumption, and could explain the enhancement of NIF activity by light under an atmosphere of H_2 (EISBRENER et al. 1978). Whatever the explanation is (see FAY 1992), these findings argue the importance of photosynthesis-respiration interplay in nitrogen-fixation by cyanobacteria.

Reversible hydrogenase is not linked to nitrogenase (HOUCHINS and BURRIS 1981c, HOUCHINS 1984) in cyanobacteria. Under anoxic growth conditions this enzyme enables the cyanobacteria to use exogenous hydrogen as reductant (HOUCHINS and BURRIS 1981c); moreover it may catalyse the production of H_2 during dark anaerobic fermentative growth of cyanobacteria.

To increase the efficiency of NIF by the reabsorbtion of the hydrogen evolved, the Hup activity must be as high as possible. Both, the affinity for hydrogen and the concentration of enzyme in the cell could be optimised concerning the above aim, as well as its coupling with the photosynthetic and respiratory chain. The increased NIF performances could be exported not only for biomass production and related application (pigments, proteins, vitamins, aminoacids, etc.), but also for the production of ammonium to be used as inorganic fertiliser in agriculture. Of course with a modified nitrogenase unable to reduce protons, it is of no interest to manage Hup for further increasing NIF. However, when it comes to hydrogen photoproduction by cyanobacteria, an environmental friendly technology, there is an opposite requirement: the lack of Hup to obtain all the hydrogen produced by nitrogenase in gas phase.

3 Applications

3.1 N_2-fixation

Nitrogen is one of the major components of the living matter. Besides the fact that the atmosphere of the Earth contains N_2 as the main component, only some cyanobacteria and other bacteria are able to use it as sole nitrogen source. The process of NIF performed by cyanobacteria either free or in symbiosis with fungi or plants (ferns, liverworts, gymnosperms and angiosperms) is very important both for natural ecosystems and agroecosystems (FOGG et al. 1973, POSTGATE 1980, 1987, SWAMINATHAN 1982); moreover, the cost of such biofertilisers is only a small fraction of that required for chemical fertilisers (RAMACHANDRAN and NATESH 1989). However, to sustain crop production, the nitrogen content of the field has to be supplemented with additional nitrogen fertilisers industrially synthesised.

The improving of NIF in cyanobacteria, as in other bacteria, either free or in symbiosis would be the background for an environmental friendly technology. The increased quantities of fixed nitrogen thus obtained could be used as fertilisers.

The following are the proposed modifications to improve the capability of cyanobacteria to fix nitrogen either in bioreactors and outdoor tanks or in natural environments:

i) Unicellular and filamentous cyanobacteria able to protect their cells against oxygen by synthesising extracellular layers and/or to temporally decrease their PS2 activity.

ii) Derepressed mutants for nitrogenase would be very useful to carry out NIF in media where some sources of fixed nitrogen are also available.

iii) Modified nitrogenase unable to reduce protons and/or having a better coupling with ferredoxin (or related carriers) would have a higher yield of NIF.

iv) Cyanobacteria either unicellular, filamentous or heterocystous with increased respiratory activity even for given period of time would have a better metabolic coupling between respiration and NIF. Mutants able to reuse molecular hydrogen evolved during NIF as well as capable of utilising exogenously added hydrogen to sustain nitrogenase activity, either via a light dependent reaction or oxy-hydrogen catabolic reaction.

v) Mutants able to synthesise new and/or higher levels of electron carriers involved in ferredoxin reduction.

vi) Considering the fact that one of the most successful living system concerning nitrogen-fixation is the symbiosis between the fern *Azolla* sp. and the cyanobacterium *Anabaena azollae* (POSTGATE 1992), it would be interesting and, probably, very profitable to use modified cyanobacteria as above (i-v) to try to improve the efficiency of nitrogen-fixation by symbiotic associations with algae, fungi, liverworts, ferns and higher plants.

3.2 N-metabolism

Nitrogen metabolism is important for environmental biotechnology also because cyanobacteria are used to produce ammonia that could be used as fertiliser instead of industrially synthesised chemicals. Ammonia production is carried out using either N_2 as sole nitrogen source or through nitrate or nitrite assimilation.

In the first case (MUSGRAVE et al. 1982, RAMOS et al. 1984), ammonia is released in the medium during NIF which can further inhibit nitrogenase activity (MAGASANIK 1977, TYLER 1978, REICH and BÖGER 1989). Modified cyanobacteria as suggested above (i-v) would be interesting to be studied for ammonia production.

In the second case, nitrate reduction is coupled with photosynthetic activity (LARA et al. 1987, ERDMAN et al. 1992) and ammonia accumulation in the medium is dependent on the inhibition of glutamine synthetase responsible for ammonia assimilation (MARTINEZ et al. 1989). Further improvements could be arrived by using cells:

i) partly impaired in ammonia assimilation (SAKHURIEVA et al. 1982).

ii) with increased nitrate reductase activity and a better coupling between nitrate reduction and photosynthetic activity.

3.3 Hydrogen production

Molecular hydrogen production is an environmental friendly biotechnology because:

i) The raw material is water.
ii) The product of hydrogen combustion is water.
iii) The source of energy is sunlight.

Cyanobacteria are able to evolve H_2 in a process catalysed by nitrogenase; this reaction is a wasteful one but under NIF conditions the hydrogen thus evolved is recycled by Hup. When the enzyme system is active but no N_2 is available (BENEMANN and WEARE 1974, WEISSMAN and BENEMANN 1977, XIANKONG et al. 1983, MARKOV et al. 1993), reducing equivalents reaching nitrogenase are used only for proton reduction. Under these conditions, it is thus possible to obtain H_2 at rates that seem promising for large scale applications (BENEMANN and WEARE 1974, HALL and RAO 1989).

However, many problems have to be solved before an economic reliable biotechnology will be accomplished. The most important are:

i) Only a minor part from the photosynthetic electron transport is directed towards nitrogenase either directly or indirectly (respiratory catabolism).

ii) Usually H_2 is obtained at low concentration mixed with other gases (argon or helium) which are furthermore very expensive.

iii) Functional life-time of the biophotolitic system is rather short.

Some of the problems related to large-scale applications of intact cyanobacteria for H_2 production could be overcome by employing some of the already proposed suggestions (i, ii, iv, and vi).

To increase H_2 production by cyanobacteria, nitrogenase and Hup have to be as follows:

a) A modified protein properly impaired in nitrogen reduction but having a high affinity for proton reduction, thereby the entire flux of reducing equivalents reaching nitrogenase will be directed towards H_2 production.

b) The absence of any Hup activity so that hydrogen evolved is not consumed, as has been done in Rhodospirillaceae (KERN et al. 1994).

3.4 Biosensors

Biosensors, a new class of analytical devices have become increasingly useful for the detection of a wide range of chemicals involved in environmental pollution as well in chemical analysis in health care and in monitoring of biotechnological processes (KARUBE et al. 1977, ARDELEAN et al. 1983, ASTON and TURNER 1984, MARGINEANU et al. 1985, HIGGINS et al. 1987).

The biosensor is a chimera comprising a biological transducer in intimate connection with an abiotic transducer. The biological component during its function generates a gradient of substances and/or energy that is converted by the abiotic transducer generally into an electric signal. The signal generated by the biological component (enzyme, organelle, cell, tissue) depends on its biochemical activity; thus any enhancement or inhibition produced by the presence of a substrate or an inhibitor could be rapidly detected. As a whole the biosensor reversibly responds to different concentrations of the analyte by increasing or decreasing its output signal.

The ability of cyanobacteria to carry out oxygenic photosynthesis has been exploited for the construction of biosensors able to detect inhibitors of photosynthesis. Many herbicides acting as water photolysis inhibitors are thus detected using intact cyanobacteria connected with an amperometric transducer (RAWSON et al. 1987,1989, RAWSON 1993). The following are some of the proposed achievements to obtain improved biosensors for environmental monitoring and biotechnology:

i) The increased respiratory activity, especially at the level of heterocysts could be exploited to develop a reliable biosensor for the pollutants acting as respiratory inhibitors; moreover the species able to fix N_2 and grow photoheterotrophically (e.g., *Anabaena variabilis*) could be used to detect carbohydrates that sustain growth and increase respiratory activity. The flexibility of cyanobacteria-based biosensors to respond to different pollutants either by a decrease (inhibitors of respiration) or an increase (organic substrates, glucose or fructose) in respiratory activity is very important for environmental monitoring. This is why such a biosensor would be a successful candidate to work under real life conditions with benefits both for the producers and the users.

ii) As it is generally accepted that in light respiration is decreased (JONES and MYERS 1963, SCHERER et al. 1988), the interplay between respiration and photosynthesis could enable us to manage one process better by changing the other one. There is no report on such a biosensor yet, but its usefulness ranges from fundamental research to practical application. Alternatively, besides there are few reports in the literature on the changing of photosynthetic rate during the transition from photoautotrophic to photoheterotrophic growth (VALIENTE et al. 1992, MANNAN and PAKRASI 1993). However, there is no report on the use of this opportunity to increase the performances of the biosensors.

iii) Cyanobacteria able to synthesise new respiratory or photosynthetic electron carriers as is the case in *Synechococcus* PCC 7942 (GEERTS et al. 1994) or to change the amount of an already functioning electron carrier could also be useful biocatalyst in (amperometric) biosensors.

3.5 Bioremediation

Cyanobacteria modified as suggested above could be used for another application related to environmental biotechnology, namely the removal of organic and/or inorganic pollutants.

3.5.1 N and P

Cyanobacteria can use nitrogen and phosphorus in the salt forms during the tertiary treatment of polluted waters, thereby decreasing their levels below accepted standards (HASHIMOTO and FURUKAWA 1989, GARBISU and HALL 1993). It could be possible that some of the above suggested changes in cyanobacteria could be fruitful for enhancing the removal of inorganic pollutants, as follows:

i) Managing the photosynthesis-respiration interplay, either to increase photosynthesis/respiration or to separate these two processes temporally/spatially, may affect their ability to utilise exogenous mineral resources.

ii) Furthermore, heterocystous cyanobacteria with or without managed respiration, photosynthesis interplay, would possess a different ability for nitrogen or phosphorous uptake.

iii) Genetically modified cyanobacteria with respect to enzymes responsible for nitrate assimilation (e.g., nitrate reductase) as well as mutants with increased levels of ferredoxin and ferredoxin reducing abilities would be particularly important for nitrogen removal from polluted waters.

3.5.2 Metal removal

Microbial mats dominated by an *Oscillatoria* sp. are effective in removing metals (Cd, Cu, Zn, Cr and Mn) from water by a mechanism involving exopolysaccharides (BENDER et al. 1989,1994). This is why cyanobacterial cells modified for their cell wall layers and/or able to oversynthesise extracellular polymers, related to NIF or not, would have higher ability to chemically bind ions than the wild cells.

3.5.3 Organic pollutant

The ability of *Anabaena* sp. strain 7120 and *Nostoc ellipsosporum* to degrade organic pollutant lindane was increased by the introduction of *linA* from *Pseudomonas paucimobilis*, the gene which product catalyses the first step of lindane catabolism (KURITZ and WOLK 1994). It could be expected that cyanobacteria engineered by the introduction of plasmids encoding the ability to metabolise other xenobiotics would be very fruitful for environmental biotechnology, specially in the case of an already modified cyanobacteria as suggested in this chapter (e.g., i-v). Moreover, a higher respiratory activity achieved either by temporal or spatial separation from photosynthesis or by increasing one respiratory electron carrier would be beneficial for the rate of organic pollutants degradation.

3.6 Biomaterials

Cyanobacteria as well as other prokaryotes synthesise and accumulate poly-3-hydroxybutyric acid (PHB) which may function as a carbon and energy storage compound, or as a sink for reducing equivalents (LIEBERGESELL et al. 1991, CAMBELL et al. 1982, VINCENZINI et al. 1990). Being non-toxic and biodegradable these compounds are already in use for the manufacture of films and bottles (BRANDL and PUCHNER 1990). Chemoorganotrophic and anoxygenic phototrophic bacteria being the most intense studied so far are the main source of PHB in biotechnology (STEINBUCHEL 1991).

However, considering the fact that in cyanobacteria PHB acts exclusively as a disposal mechanism to eliminate excess reducing equivalents (PHILIPPIS et al. 1992), cyanobacteria having an excess of ATP and NADPH could be successful candidates for biotechnological PHB production. The increased levels of available ATP and NADPH may be the result of either:

i) better temporal and spatial co-operation between respiration and photosynthesis; or

ii) decreased biosynthetic requirements generated by N or P starvation under photoauto- and photoheterotrophic conditions.

These are only few suggestions; however much work must be carried out to accomplish a profitable and environmentally safe biotechnology. Moreover, other ideas are developing all over the world.

References

Ardelean I, Margineanu DG, Vais H (1983) Electrochemical conversion in biofuel cells using *Clostridium butyrycum* or *Staphylococcus aureus*. Bioelectrochem Bioenerg 11:273-277

Aston WJ, Turner APF (1984) Biosensors and biofuel cells. Biotechnol Genet Eng Rev 1:89-120

Bender J, Archibold ER, Ibeanusi V, Gould JP (1989) Lead removal from contaminated water by a mixed microbial ecosystem. Water Sci Technol 21:1661-1665

Bender J, Rodriguez-Eaton S, Ekanemesang UM, Philipps P (1994) Characterization of metal-binding bioflocculants produced by the cyanobacterial component of mixed microbial mats. Appl Environ Microbiol 7:2311-2315

Benemann JR, Weare NM (1974) Hydrogen evolution by nitrogen-fixing *Anabaena cylindrica* cultures. Science 184:174-175

Bergman B, Siddiqui PJA, Carpentier EJ, Peschek GA (1993) Cytochrome oxidase: subcellular distribution and relationship to nitrogenase expression in the non-heterocystous marine cyanobacterium *Trichodesmium thiebautii*. Appl Environ Microbiol 59:3239-3244

Bothe H, Neuer G, Kalbe I, Eisbrener G (1982) Electron donor and hydrogenase in nitrogen fixing microorganisms. In: Stewart WDP, Gallon JR (eds) Nitrogen fixation. Academic Press Ltd, London, pp 83-112

Bradley S, Carr NG (1971) The absence of functional photosystem II in heterocysts of *Anabaena cylindrica*. J Gen Microbiol 68:xiii-xiv

Brandl H, Puchner P (1990) The degradation of shampoo bottles in a lake. In: Dowes EA (ed) Novel biodegradable microbial polymers. Kluwer Academic Publishers, Dordrecht, pp 421-422

Bryant DA (ed) (1994) The molecular biology of cyanobacteria. Kluwer Academic Publishers, Dordrecht

Burris RH (1991) Nitrogenases. J Biol Chem 266:9336-9342

Cambell JIII, Sevens SE, Blakwill DL (1982) Accumulation of poly-β-hydroxybutyrate in *Spirulina platensis*. J Bacteriol 149:361-363

Carr NG, Whitton BA (eds) (1973) The biology of blue-green algae. Blackwell Scientific Publications Ltd, Oxford

Dews G (ed) (1990) The molecular biology of membrane-bound complexes in phototrophic bacteria. Plenum, New York

Dixon ROD (1972) Hydrogenase in legume root nodule bacteroids: occurrence and properties. Arch Microbiol 85:193-201

Donze M, Haveman J, Schiereck P (1972) Absence of photosystem 2 in heterocysts of the blue-green alga *Anabaena*. Biochem Biophys Acta 256:157-161

Eisbrener G, Distler E, Floener L., Bothe H (1978) The occurrence of hydrogenase in some blue-green algae. Arch. Microbiol 118:174-184

Erdman N, Zuther E, Abarzua S (1992) Comparative studies on the photoproduction of non-hydrogenous resources by cyanobacteria. Curr Microbiol 25:83-87

Fay P (1992) Oxygen relations in nitrogen fixation in cyanobacteria. Microbiol Rev 56:340-373

Fay P, van Baalen C (eds) (1987) The cyanobacteria. Elsevier BV, Amsterdam

Fogg GE, Stewart WDP, Fay P, Walsby AE (eds) (1973) The blue-green algae. Academic Press, London

Gallon JR, LaRue TA, Kurtz GW (1974) Photosynthesis and nitrogenase activity in the blue-green alga *Gloeocapsa*. Can J Microbiol 20:1633-1636

Garbisu C, Hall DO (1993) Removal of phosphate by foam-immobilised *Phormidium laminosum* in batch and continuous-flow bioreactor. J Chem Tech Biotechnol 57:181-189

Geerts D, Schubert H, de Vrieze G, Borrias M, Matthijs HCP, Weisbeek PJ (1994) Expression of *Anabaena* PCC 7937 plastocyanin in *Synechococcus* PCC 7942 enhances photosynthetic electron transfer and alters the electron distribution between photosystem I and cytochrome-c oxidase. J Biol Chem 269:28068-28075

Hall DO, Rao KK (1989) Immobilised photosynthetic membranes and cells for the production of fuels and chemicals. Chim Oggi 7:41-47

Hashimoto S, Furukawa K (1989) Nutrient removal from secondary effluent by filamentous algae. J Ferment Bioengn 67:62-69

Higgins IJ, Swain AS, Turner APF (1989) Principles and applications of biosensors in microbiology. J Appl Bacteriol Symposium Supplement 93S-104S

Houchins JP (1984) The physiology and biochemistry of hydrogen metabolism in cyanobacteria. Biochim Biophys Acta 768:227-255

Houchins JP, Burris RH (1981a) Light and dark reactions of uptake hydrogenase in *Anabaena* 7120. Plant Physiol 68:712-716

Houchins JP, Burris RH (1981b) Physiological reactions of the reversible hydrogenase from *Anabaena* 7120. Plant Physiol 68:717-721

Houchins JP, Burris RH (1981c) Occurrence and the localization of two distinct hydrogenases in the heterocystous cyanobacterium *Anabaena* sp. strain 7120. J Bacteriol 146:209-214

Houchins JP, Hind G (1983) Flash spectroscopic characterization of photosynthetic electron transport in isolated heterocysts. Arch Biochem Biophys 224:272-282

Jones LW, Myers J (1963) A common link between photosynthesis and respiration in blue-green algae. Nature 199:670-672

Karube I, Matsunaga T, Mitsuda S, Suzuki S (1977) Microbial electrodeBOD sensors. Biotechnol Bioengn 19:1535-1548

Kerby NW, Musgrave SC, Rowell P, Shestakov SV, Stewart WDP (1986) Photoproduction of ammonia by immobilized mutant strains of *Anabaena variabilis*. Appl Microbiol Biotechnol 24:42-46

Kern M, Klipp W, Klemme JH (1994) Increased nitrogenase-dependent H_2 photoproduction by *hup* mutants of *Rhodospirillum rubrum*. Appl Environ Microbiol 60:1768-1744

Kumazawa S, Mitsui A (1992) Photosynthetic activity of a synchronously grown aerobic N_2-fixing cyanobacterium *Synechococcus* sp. Miami BG 043511. J Gen Microbiol 136:467-472

Kuritz T, Wolk CP (1994) Use of filamentous cyanobacteria for biodegradation of organic pollutants. Environ Microbiol 61:234-238

Lara C, Romero JM, Coronil T, Guerrero MG (1987) Interaction between photosynthetic nitrate assimilation and CO_2 fixation in cyanobacteria. In: Ullrich WR, Aparicio PJ, Syrett PJ, Castillo F (eds) Inorganic nitrogen metabolism. Springer Verlag, Berlin-Heidelberg, pp 45-53

Leon C, Kumazawa S, Mitsui A (1986) Cyclic appearance of aerobic nitrogenase activity during synchronous growth of unicellular cyanobacteria. Curr Microbiol 13:149-153

Liebergesell M, Hustede E, Timm A, Steinbuchel A, Fuller RC (1991) Formation of poly (3-hydroxyalkanoates) by phototrophic and chemoheterotrophic bacteria. Arch Microbiol 155:415-421

Lindblad P, Sellstedt A (1990) Occurrence and localization of an uptake hydrogenase in the filamentous heterocystous cyanobacterium *Nostoc* PCC 73102. Protoplasma 159:9-15

Magasanik B (1977) Regulation of bacterial nitrogen assimilation by glutamine synthetase. Trends Biochem Sci 2:9-12

Mannan MR, Pakrasi HB (1993) Dark heterotrophic growth conditions result in an increase in the content of photosystem II units in the filamentous cyanobacterium *Anabaena variabilis* ATCC 29413. Plant Physiol 103:971-977

Margineanu DG, Vais H, Ardelean I (1985) Bioselective electrodes with immobilized bacteria. J Biotechnol 3:1-9

Markov SA, Lichtl R, Rao KK, Hall DO (1993) A hallow fibre photobioreactor for continuous production of hydrogen by immobilized cyanobacteria under partial vacuum. Int J Hydrogen Energy 18:901-906

Martinez A, Llama MJ, Alaña A, Serra JL (1989) Sustained photoproduction of ammonia from nitrate or nitrite by permeabilized cells of the cyanobacterium *Phormidium laminosum*. J Photochem Photobiol 3:269-279

McCarthy JEG, Ferguson SJ, (1982) Respiratory control and the basis of light-induced inhibition of respiration in chromatophores from *Rhodopseudomonas capsulata*. Biochem Biophys Res Commun 107:1406-1411

Mortenson LE (1978) The role of dihydrogen and hydrogenase in nitrogen fixation. Biochimie 60:219-224

Mullineaux PM, Gallon JR, Chaplin AE (1981) Acetylene reduction (nitrogen fixation) by cyanobacteria grown under light-dark cycles. FEMS Microbiol Lett 10:245-247

Murry MA, Wolk CP (1989) Evidence that the barrier to the penetration of oxygen into heterocysts depends upon two layers of the cell envelope. Arch Microbiol 151:469-474

Musgrave SC, Kerby NG, Codd GA, Stewart WDP (1982) Sustained ammonia production by immobilized filaments of the nitrogen-fixing cyanobacterium *Anabaena* 27893. Biotechnol Lett 4:647-652

Packer L, Glazer AN (eds) (1988) Methods in enzymology, vol 167. The Cyanobacteria. Academic Press Inc, San Diego

Paerl HW, Prufert LE, Ambrose WW (1991) Contemporaneous N_2-fixation and oxygenic photosynthesis in the non-heterocystous mat-forming *Lyngbya aestuarii*. Appl Environ Microbiol 57:3086-3092

Peterson RB, Burris RH (1978) Hydrogen metabolism in isolated heterocysts of *Anabaena* 7120. Arch Microbiol 116:125-132

Philippis De R, Claudio S, Vincenzini M (1992) Glycogen and poly-β-hydroxybutyrate synthesis in *Spirulina maxima*. J Gen Microbiol 138:1623-1628

Postgate JR (1980) The nitrogen economy of marine and land environments. In: Blaxter K (ed) Food chains and human nutrition. Applied Science Publishers, London, pp 161-181

Postgate JR (1987) Prospects for the improvement of biological nitrogen fixation. J Appl Bacteriol Symposium Supplement 85S-91S

Postgate JR (ed) (1992) Microbes and man, Third edition, Cambridge University Press, Cambridge

Ramachandran S, Natesh S, (1989) Agriculture and biotechnology-challenges and opportunities in India. In: Leone U, Rialdi G, Vanore R (eds) International workshop on Advanced technologies for increased agricultural productions. Stampa Basile, pp 12-22

Ramos JL, Guerrero MG, Losada M (1984) Sustained photoproduction of ammonia from dinitrogen and water by the nitrogen-fixing cyanobacterium *Anabaena* strain ATCC 33047. Appl Environ Microbiol 48:114-118

Rawson DM (1993) Bioprobes and biosensors. In: Calow, P (ed) Handbook of ecotoxicology, vol 1. Blackwell Scientific Publications, Oxford, pp 428- 438

Rawson DM, Willmer AJ, Cardossi MF (1987) The development of whole cell biosensors for on-line screening of herbicide pollution of surface waters. Toxicity Assessments 2:325-340

Rawson DM, Willmer AJ, Turner APF (1989) Whole-cell biosensors for environmental monitoring. Biosensors 4:299-311

Reich S, Böger P (1989) Regulation of nitrogenase activity in *Anabaena variabilis* by modification of the Fe protein. FEMS Microbiol Lett 58:81-86

Rugolo M, Zannoni D, (1983) oxygen-induced inhibition and light -dependent uptake of tetraphenylphosphonium ions as a probe of direct interaction between photosynthetic and respiratory components in cells of *Rhodopseudomonas capsulata*. Biochem Biophys Res Commun 113:155-162

Sakhurieva GN, Polukhina LE, Shestakov SV (1982) Glutamine synthesis in *Anabaena variabilis* mutants with derepressed nitrogenase. Microbiol 51:308-312

Scherer S, Almon H, Böger P (1988) Interaction of photosynthesis, respiration and nitrogen fixation in cyanobacteria. Photosyn Res 15:95-114

Shelly DC, Vermaas WJF (1989) Directed mutagenesis to probe structure and function of photosystem II. Physiol Plant 77:436-443

Stal LJ, Heyer H, Jacob G (1990) Occurrence and role of poly hydroxyalkanoates in the cyanobacterium *Oscillatoria limosa*. In: Novel biodegradable microbial polymers. Kluwer, Dordrecht, pp 435-438

Stal LJ, Krumbein WE (1981) Aerobic nitrogen fixation in pure cultures of the benthic marine *Oscillatoria* (cyanobacteria). FEMS Microbiol Lett 11:295-298

Stal LJ, Krumbein WE (1987) Temporal separation of nitrogen fixation and photosynthesis in the filamentous non-heterocystous cyanobacterium *Oscillatoria* sp. Arch Microbiol 149:76-80

Stanier RY, Sistrom WR, Hansen TA, Whitton BA, Castenholz RW, Phennig N, Gorlenko VN, Kondratieva EN, Eimijellen KE, Whittenburry R, Cerna RL, Truper HG (1978) Proposal to place the nomenclature of cyanobacteria (blue-green algae) under the rules of International Code of Nomenclature of Bacteria. Internatl J Systematic Bacteriol 28:335-336

Steinbuchel A, (1991) 3-polyhydroxyalkanoic acids. In: Byrom D (ed) Biomaterials, novel materials from biological sources. Macmilan Publishers Ltd, Bosingstake, pp 123-213

Stewart WDP, Gallon JR (ed) (1982) Nitrogen fixation. Academic Press Ltd, London

Stewart WDP, Rowell P, Tel-Or E (1975) Nitrogen fixation and the heterocysts in blue green algae. Biochem Soc Trans 3:357-361

Swaminathan MS (1978) Biotechnology research in the third world agriculture. Science 218:967-972

Tel-Or E, Stewart WDP (1975) Manganese and photosynthetic oxygen evolution by algae. Nature 258: 715-716

Tyler B (1978) Regulation of assimilation of nitrogen compounds. Annu Rev Biochem 47:1127-1162

Valiente EE, Nieva M, Avendaño C, Maesao ES (1992) Uptake and utilization of fructose by *Anabaena variabilis* ATCC 29413. Effect on respiration and photosynthesis. Plant Cell Physiol 33:307-313

Van Niel CB (1941) The bacterial photosynthesis and their importance for the general problem of photosynthesis. Adv Enzymol 1:263-328

Vermaas WJF, Charite J, Shen G (1990) Q_A binding to D_2 contributes to the functional and structural integrity of photosystem II. Z Naturforsch 450:359-365

Vincenzini M, Sili C, Philippis de R, Ena A, Materassi R (1990) Occurrence of poly-β-hydroxybutyrate in *Spirulina* species. J Bacteriol 172:2791-2792

Weissman JC, Benemann JR (1977) Hydrogen production by nitrogen-starved cultures of *Anabaena cylindrica*. Appl Environ Microbiol 33:123-131

Xiankong Z, Haskell JB, Tabita FR, van Baalen C (1983) Aerobic hydrogen production by heterocystous cyanobacteria *Anabaena* spp. strains CA and 1F. J Bacteriol 156:1118-1122

Author Index

D

T

W

X

Y

Organism Index

GPSR Compliance

The European Union's (EU) General Product Safety Regulation (GPSR) is a set of rules that requires consumer products to be safe and our obligations to ensure this.

If you have any concerns about our products, you can contact us on ProductSafety@springernature.com

In case Publisher is established outside the EU, the EU authorized representative is:

Springer Nature Customer Service Center GmbH
Europaplatz 3
69115 Heidelberg, Germany

Batch number: 10755284

Printed by Printforce, the Netherlands